Riding the Wave

Emergency managers and public safety professionals are more frequently being called on to address increasingly challenging and complex critical incidents, with a wider variety and intensity of hazards, threats, and community vulnerabilities. Much of the work that falls into the scope of emergency managers – prevention, preparedness, mitigation – is "blue sky planning" and can be contained and effectively managed within projects. This book provides a foundational project management methodology relevant to emergency management practice and explains and demonstrates how project management can be applied in the context of emergency and public safety organizations.

Special features include:

- an initial focus on risk assessment and identification of mitigation and response planning measures;
- a clear set of better practices, using a diverse set of examples relevant to today's emergency environment, from projects to develop emergency response exercises to application development to hazard mitigation;
- a framework for managing projects at a strategic level and how to incorporate this into an organization's program, as well as how to develop and manage an emergency program and project portfolio; and
- suitability as both a hands-on training guide for emergency management programs and a textbook for academic emergency management programs.

This book is intended for emergency managers and public safety professionals who are responsible for developing emergency programs and plans, including training courses, job aids, computer applications and new technology, developing exercises, and for implementing these plans and components in response to an emergency event. This audience includes managers in emergency and first response functions such as fire protection, law enforcement and public safety, emergency medical services, public health and healthcare, sanitation, public works, business continuity managers, crisis managers, and all managers in emergency support functions as described by FEMA. This would include those who have responsibility for emergency management functions, even without the related title.

Andrew Boyarsky, MSM, PMP, CBCP, cABCF, is President of Pinnacle Performance Management, and an emergency management and disaster recovery specialist with 30 years of experience in project management and 23 years in emergency management, business continuity, and disaster recovery. He is Clinical Associate Professor and teaches at NYU, John Jay College of Criminal Justice, and at Yeshiva University. In the early 1990s, he developed and managed large-scale emergency medical and mass care response projects overseas in the former Yugoslavia and in the Caucasus with the International Federation of the Red Cross and Red Crescent Societies and Catholic Relief Services/ Caritas Internationalis. From 2007 to 2016 Boyarsky was Project Manager for Coastal Storm Plan Training working on behalf of the NYC Office of Emergency Management, responsible for training over 30,000 city staff for emergency sheltering operations. This included managing additional training programs in logistics, recovery, special medical needs, and disability, access, functional needs, and pandemic response.

His clients have included FEMA, the Regional Catastrophic Planning Team of NY-NJ-CT-PA, New York City Office of Emergency Management, New York City Health Department, Los Angeles Emergency Management Department, Baltimore City Health Department, Westchester County (NY) Office of Emergency Management, and the NYC Human Resources Administration.

He has a B.A. from Johns Hopkins University, Masters in the Science of Management from the Hult International School of Business (formerly the Arthur D. Little School of Management), and earned his PMP certification in 2003, CBCP in 2019, and cABCF in 2020. He also hosts a podcast series, Riding the Wave: Project Management for Emergency Managers, and volunteers with the Community Emergency Response Team in South Orange, NJ, where he resides with his family.

SECURITY, AUDIT AND LEADERSHIP SERIES

Series Editor: Dan Swanson, Dan Swanson and Associates, Ltd., Winnipeg, Manitoba, Canada.

The *Security, Audit and Leadership Series* publishes leading-edge books on critical subjects facing security and audit executives as well as business leaders. Key topics addressed include Leadership, Cybersecurity, Security Leadership, Privacy, Strategic Risk Management, Auditing IT, Audit Management and Leadership

Riding the Wave

Applying Project Management Science in the Field of Emergency Management

Andrew Boyarsky

CRC Press
Taylor & Francis Group
Boca Raton London New York

CRC Press is an imprint of the
Taylor & Francis Group, an **informa** business

Designed cover image: ©Shutterstock

First edition published 2024
by CRC Press
2385 NW Executive Center Drive, Suite 320, Boca Raton FL 33431

and by CRC Press
4 Park Square, Milton Park, Abingdon, Oxon, OX14 4RN

CRC Press is an imprint of Taylor & Francis Group, LLC

© 2024 Andrew Boyarsky

ISBN: 9781032062853 (hbk)
ISBN: 9781032062860 (pbk)
ISBN: 9781003201557 (ebk)

DOI: 10.1201/9781003201557

Typeset in Sabon
by Newgen Publishing UK

Contents

PART 3
Strategic Project Management

Preface

When the situation was manageable it was neglected, and now that it is thoroughly out of hand, we apply too late the remedies which then might have effected a cure. There is nothing new in the story. It is as old as the Sibylline books. It falls into that long, dismal catalogue of the fruitlessness of experience and the confirmed unteachability of mankind. Want of foresight, unwillingness to act when action would be simple and effective, lack of clear thinking, confusion of counsel until the emergency comes, until self-preservation strikes its jarring gong—these are the features which constitute the endless repetition of history.

Winston Churchill, House of Commons, May 2, 1935

We live in a VUCA[1] world, one that is filled with volatility, uncertainty, complexity, and with a sense of ambiguity as to where we may be headed, whether on a local or global community level. Let's face it, you don't need to look too far to see these signs that are creating this environment: natural hazards, human-caused threats, social and political strife, technological complexities, and financial and economic instability. Just read the news headlines to witness the manifestations. There are wildfires in the Western North America and Canada, widespread flooding in the Midwest and South Asia, increasing intensity of tropical cyclones, and tornadic activity spreading well beyond its typical range.

Roughly 12 years ago I had the opportunity to work on an environmental literacy course for the City University of New York. While conducting research for the course, one of the key points was the idea of the "Tipping Point".[2] This is where we move beyond the "Point of No Return" with increased global warming into runaway climate change with very unstable and extreme weather patterns. The very sad truth is that we are now beyond that tipping point. While the alarm bells were sounding loud and clear over a decade ago, the situation has worsened, with no signs of any improvement. The recent report by the Intergovernmental Panel on Climate Change (IPCC)[3] indicates that we are 1°C (roughly 1.8°F) above pre-industrial levels and will arrive at 1.5°C by 2030. This is a little more than ten years away as of this writing. Yet, we are already witnessing these extreme weather phenomena.

Being in the middle of this emerging phenomena makes it challenging at best to understand what is happening. What we commonly refer to as climate change is certainly having a destabilizing effect on weather patterns and consequent impacts on our environment. Human settlement and consumption are exacerbating the situation, not only in the green-house gases that are emitted, but in the ways in which we interact with our environment. Here are just some examples:

- An extended wildfire season, moving from five months in the 1970s to now year-round; increased temperatures are causing severe drought, melting mountain snow packs earlier and leaving more areas exposed to burning. The Marshall Fire in Boulder, Colorado, in December 2021 is a clear example of this, causing over $500 million in damage to a major suburban community.[4]
- Increasing ocean water temperatures lead to greater evaporation and conditions that spawn and sustain larger storms (hurricanes/typhoons), monsoons, rainstorms, microbursts, etc.; this is true not only in the Western hemisphere but also in the Eastern hemisphere – witness the massive flooding in South Asia, displacing many millions of inhabitants. Similar flooding has been impacting the United States: the Mississippi River and Tennessee Valley in the past decade, just to name a few examples.
- Equally disturbing, increasing ocean temperatures are disrupting ecosystems, reducing ocean plant and animal life; corals are disappearing at an alarming rate and house numerous coastal fish populations.
- Season creep, wherein plants bloom earlier causing some species to come out earlier, leaves other species unable to compensate (what might be called a "late to dinner" effect). This has led to some animal populations beginning to collapse: migratory birds, insects, deer.
- Increasing development and hardscaping along river fronts and coastal areas have reduced natural watersheds and increased the extent and severity of flooding levels. As a clear example of this, Florida once was all wetlands, a giant peninsula of marshes, swamp, and grasslands, with a healthy ecosystem. We know the story; the swamp was dredged, water management put in place, and what we know as the Sun and Gold coasts, filled with resort communities, retirement homes and buildings, golf courses, etc. grew exponentially. But Mother Nature has a way of taking back what is rightfully hers.
- A similar trend is happening in the forested and mountain areas of the western United States; housing developments are pushing further into forested areas (Wildland Urban Interface or WUI), increasing their risk exposures to wildfires. While the doctrine of fire suppression by the US Fire Service was called into question and the FLAME Act was designed to address the shortage in resources in battling wildfires, we still see development unabated in western US states.
- Encroachment into corners of the globe where humans consume and trade in wild animals and comingling species where viruses can transmit and mix to foster greater potential for pathogens which can threaten human lives. The spread of Ebola and Monkeypox are the clearest examples of this type of pathogenic spread.

Our political structures are currently ill equipped to fully address the scope and scale of these threats and impacts to the environment. While some political leaders at the national and regional levels are tackling this challenge head on (a good example is the 100 resilient cities program[5]), other political leaders are reluctant to take the dramatic steps to reduce our common carbon footprint and, furthermore, to invest what is necessary in the immediate future for robust mitigation measures against natural hazards, such as flood barriers/

sea walls, levees, storm cellars (for tornados), seismic retrofitting (for earthquake), and forest maintenance, and establishing tougher building codes and zoning restrictions. Many of these measures are unpopular, as they do not show any clear, immediate "Return on Investment" and require political will and fortitude, moreover, and broad social support in the face of business development interests, which is why these efforts are often stymied.

We can see that our legal frameworks are strained and, at times, frayed by the advent of new technologies and perils, and our failure to harness them through legal and regulatory means: blockchain (e.g. unregulated cross-border currencies such as Bitcoin), the dark web, and the Internet of Things (IoT), just to name a few. On the social front, our natural, local community structures, supported by government, community organizations, and faith-based communities are weakened as our "24hrs./365 day/year, always-on" culture leaves little time for us to attend to immediate, long-term local issues and to connect across traditional local community and faith-based groups. With an increasingly transient society, it is challenging to muster a sense of community in order to gain consensus around concrete measures to deal with the impacts of climate change. Civil society has been further balkanized by the forces of political extremism and proliferation of social media as a primary source of news,[6] where groups create narrow, echo chambers for their own lines of thought without a strong consensus of what we see as a "common operating picture" of the potential challenges we face.

We are in great need of substantial measures along the panoply of risk responses to address the potential impacts of the threats and hazards that we face. In the absence of meaningful prevention and mitigation measures to eliminate or reduce vulnerability to risk impacts, professionals in emergency management and public safety will be called upon to address these perils and their consequences to protect and safeguard our communities.

NOTES

1 VUCA is believed to have been developed as a term at the US Army War College and is based on the management theories of Warren Benis and Burt Nanus. See: http://usawc.libanswers.com/faq/84869
2 Beyond the Point of No Return, December 12, 2007, See: www.heatisonline.org/contentserver/objecthandlers/index.cfm?ID=6752&method=full#_edn1; for a short video on the subject, see: http://wakeupfreakout.org/film/tipping.html
3 See www.ipcc.ch/sr15/chapter/summary-for-policy-makers/
4 As many as 600 homes lost, six people injured as Marshall fire quickly spreads across Boulder County. The Colorado Sun. Boulder, Colorado. December 30, 2021. Retrieved January 1, 2022. https://coloradosun.com/2021/12/30/boulder-grass-fire-evacuations/
5 See www.100resilientcities.org/
6 Experts Say the 'New Normal' in 2025 will be Far More Tech-Driven, Presenting More Big Challenges, *Pew Research Center, Internet and Technology*, February 18, 2021 www.pewresearch.org/internet/2021/02/18/emerging-change/

Acknowledgments

There is a saying that when we have accomplished something that appears to stand out from the collective work of our peers that we are "standing on the shoulders of those who went before us". I have said that many times but was never quite comfortable with the analogy. I would prefer to say that those who went before me helped blaze the trail, and, that together, we have arrived at a higher level of understanding, so, with that made clear, I want to thank the following people who have helped to guide me and make this book possible.

First and foremost, I want to thank my wife, Claire, who provided never ending, patient support and encouragement during the journey of writing this book. Sometimes we need to wander in the woods a bit to find our way, and, knowing that we have a guiding light brings us to the end of our journey and safely back home.

I want to thank Charles Jennings who has provided feedback and the opportunity to create elements of the book through my teaching in the Masters in Emergency Management at John Jay College. I want to express special appreciation for the time and attention of Tom Carey, Monmouth University, who provided very thoughtful and insightful edits and pointed out additional research sources.

I appreciate the help of the following colleagues for their feedback on book content and early drafts:

James Vasquez	Rick LaValla	Kelly McKinney
Paul Kearns	Steve Davis	Jeff Schlegelmilch
Bill Pappas	Gene Komondor	Todd DeVoe
Michael Schultz	Vincent Davis	James Muller
Thomas Croall	Celia Seigerman-Levit	Victor Orellana Acuña
David Lindstedt	Stuart Syme	Josh DiVicenzo
Mark Armour	Andy McGuire	

For their time in interviews for the book, many thanks go to Greg Brunelle, Erik Gaul, Justin Kates, Heather Roiter, Jason Jenkins, Andrew Phelps, and Damon Coppola. To Michael Stone, President of the IFRC Alumni Group, who has graciously permitted me to reference his example of leadership in the book.

For their feedback and edits on the narrative, Jim Lear, Dean Kameros, Emily Kameros, and Steve Weber, with the support of Stephanie Bourgeois, Yael Sivi, and Yosh Beier; my hometown crew who were ever supportive through the dark days of COVID, and put up with my Cassandra-like commentary during the pandemic.

To my publishing team, first and foremost Gabriella Williams at Taylor and Francis for her patient support, and, Radhika Gupta, editor, for their astute and artful edits, and, most importantly, Dan Swanson for his tenacity and enthusiasm, as well as his collective of authors who provided initial feedback on the early draft.

For anyone who I may have left off, I want to thank you and apologize for any oversight. I am still on a continuing journey to improve.

Part I

Setting Up the Basecamp for Projects

Chapter 1

The Accidental Project Manager

BAPTISM BY FIRE – FORMER YUGOSLAVIA AND THE MAP

In March of 1993 I landed on the ground in Zagreb, the national capital of Croatia. Croatia and neighboring Bosnia-Herzegovina were embroiled in brutal and bloody civil wars that had ensued after the break-up of Yugoslavia during 1991–1992. I was placed in charge of a medical assistance project to provide support for the medical needs of refugees and displaced persons in Croatia. Before my arrival, I had spent three days of orientation at the organization's headquarters in Baltimore, MD[1] and had read up prior to that on the Balkans' history, culture, and what led up to the war. Armed with this and an undergraduate degree in international relations, as well as some experience working in Eastern Europe, I felt ready to take on this new emergency humanitarian mission.

My taxi pulled up to the low whitewashed building just off the square of the main cathedral in Zagreb, where I emptied myself and all my baggage into the vestibule. At the door I was greeted by Sister Annamaria Šimić,[2] a petite, smiling nun who was born in Northern Bosnia. While short of stature, she made up for it in energy and fortitude. Once I had a chance to get settled in my hotel, our initial task at hand would be to look for office space.

I would not even get a chance to get my bags unpacked when I was immediately presented with my first urgent request; there were two refugees in the hospital with leukemia, whose doctors had submitted a list of needed medicines for their treatment. My project was not even operational, and I had a decision thrust upon me. The total cost between these two would have been about $60,000, a significant chunk of the $3.6 million budget I had to spend on pharmaceuticals. If I did agree to cover the cost, then at this rate of spending the project might wind up quickly exhausting the budget while covering a small number of needs. If I did nothing and the medicines were not provided, the odds were that these two young people might not survive.

That evening I called the US Embassy since the US State Department was the funding agent for the aid grant.[3] I figured that someone there could offer some guidance with this situation. It was the weekend, and my call was transferred to a duty officer who, after asking me the particulars of the patient cases, said there was little that he could do to help and then offered little encouragement by adding "I wouldn't want to be in your shoes". "Thanks" I thought.

What I did not realize at the time was that I was presented with a problem of project scope. If I decided to go the route of helping the few with acute medical needs such as

DOI: 10.1201/9781003201557-2

leukemia, then the project would have a limited reach in beneficiary numbers and most likely a shorter duration and geographical reach due to the limited number of hospitals in the country. If, on the other hand, I purchased inexpensive medicines and medical supplies, then the project might reach a broader population of refugees and displaced persons in the country. However, quick calculation of the average dollar per beneficiary would yield roughly $7.2 per beneficiary, a piddling amount even in war-torn Croatia. The solution clearly was somewhere in between the two options if the project was to have any significant impact.

Whatever course of action I chose, I still needed to come to a decision. Not making a decision was still a decision, with consequences attached. I decided to seek a private donation through my organization and asked the hospital to describe the two cases, respecting the privacy of the two patients, and submitted this along with an urgent request with a list of specific medicines and approximate cost to our headquarters. This request was replicated to the broader network of charities we were affiliated with, and in due time a donation was made for one of the cases; one patient had a positive prognosis, while the other's was negative and, unfortunately, was unlikely to survive. This was also among one of my first lessons; the project was not going to save all of the targeted beneficiaries, and not every outcome was going to be an unmitigated success. In a major disaster situation, including the impacts of a terrible civil war, the mission of the emergency manager is to do the greatest good for the largest number of people.

Amid all of the project documentation I had lugged around with me was a six-page project plan and a stack of organizational and US Government grant regulations and guidelines that stood about a foot high. The six-page project plan lay on the right side of the desk and provided no practical guidance as to how to clearly execute the project. The documentation on the left side of the desk appeared rather daunting and unlikely to yield anything helpful. Upon seeking advice from the head office as to what I should do, I was told: "read through all of the regulations and the project plan and decide on the appropriate course of action". After reading the first ten pages of the government regulations and realizing there was scant guidance as to my specific project or the operational environment—a civil war zone—I decided my logical course of action would be to use the six-page project proposal as my general guidance and my common sense to plan and execute the project. When I had time later on, I would read through the regulations for any relevant guidance. This proved to be the wisest choice. Expediency and practicality trumped administrative bureaucracy.

In those first few days, I realized that I had to have a better gauge of the scope of my project and settled on the idea of assessing the medical needs of the refugees and displaced persons in Croatia. My first stop was with the Ministry of Health where I met with the Deputy Minister of Health. His first question that he asked was "What can I get for you?". In my youthful haste, I rattled off a list of data points that I needed: locations of high-density refugee and displaced populations (DP), refugee and DP camps, clinics and medical stations serving mostly refugees and displaced persons,[4] health surveillance data, morbidity and mortality data, etc. He waited until I was done and then politely asked with a smile "I meant coffee, tea, or juice". I quickly realized that I had a lot to learn about my communication style and the culture of the country I was in. This was major lesson number two; I would have to adapt to the local working environment and the culture. This was particularly complicated by the fact that there were many regional differences, as well as political and social considerations, not the least of which was that, quite frankly, I was working in a war zone.

The truth became apparent in that first meeting with the Ministry of Health that, in the middle of a civil war, they did not have the exact data I needed. They did have an idea of where most refugee and displaced persons were located, in encampments, makeshift housing, and hotels along the coast.[5] This information, coupled with some early meetings with representatives of the UN High Commissioner for Refugees, the World Health Organization (WHO), United Nations Children's Fund (UNICEF), and a number of non-governmental organizations (NGOs) involved in supporting the health needs of the refugee and displaced persons, created a map that set me off on a journey around the country to survey these needs in greater depth, establish working relationships and agreements to provide for these needs as a basis for the project.

Years later I would realize that it was in those early days and months that I was learning the basics of project management, learning by doing, or experiential learning as it is called today. I would gain an understanding of the basic concepts of scope, team building, monitoring and control, stakeholder management, and other skills, if not in terms of the definitions and methodologies of project management according to the Project Management Body of Knowledge (PMBOK®),[6] then on an experiential basis. It was later when I earned my Project Management Professional (PMP®) certification that I would more formally internalize this learning.

MANAGING EMERGENCY RESPONSE PROGRAMS IN EASTERN EUROPE AND PREPAREDNESS PROGRAMS FOR NYC AND THE NY/NJ REGION; SHORT LESSONS LEARNED[5]

One of the major lessons learned—practically by accident as a result of not being fully operational until months after the start—was the benefit of starting small and then expanding the project outward. Starting on a smaller scale allows for testing the system, in this case the procurement and delivery mechanism for supplies, before scaling up. This acts as a "proof of concept", so you can figure out what works and, if there are faults in the system, then better to identify those cracks in the system before they are magnified on a larger scale.

There were important considerations that forced this approach; we did not have a good grasp of the needs, had a limited number of staff, and, most importantly, I did not have a good idea of the operating environment, the hazards we might face, or our logistical resource constraints. The assessment of the medical needs, the pipeline for delivery, and logistics requirements would run parallel to that first procurement and delivery.

Three major challenges in this project all had to be balanced against one another: assessing the medical needs of refugees and displaced persons, getting essential medicines and medical supplies to those in need, and tracking those supplies from the warehouse to the end-user. What I practiced then (and what I did not know at the time) is referred to in project management as the "rolling wave approach". This means, like a rolling wave approaching, you prepare based on what you currently know and the relevant immediate near future, and so I prepared for that first wave of the project, in this instance it was the first three months. So, here is what I knew within the first two weeks:

- In March of 1993, one out of every ten people in Croatia was a refugee/displaced person, with a refugee and displaced person population of over 500,000 people in a country of about 5 million people. There were several large refugee camps in Slavonia (Eastern Croatia, near the Serbian border) and in many of the major hotels along the Dalmatian coast.
- We had a main warehouse in the capital Zagreb and three regional warehouses (Zadar, Split, and Đakovo), plus four small Sport Utility Vehicles (SUVs) (Suzuki Samurai model[7]).
- The project had nine distribution and monitoring staff who were responsible for ensuring delivery and tracking medical supplies to the medical facility (hospital, clinics, etc.), a pharmacist, and a project admin. Many of them had been hired or selected before I arrived.
- We had standing procurement contracts with several international vendors for pharmaceuticals.

Given this situation our initial major objectives (what would now be called an incident action plan [IAP] in emergency management speak) were as follows:

- Establish a logistics system including delivery mechanism, tracking, and reporting, and train staff on this nation-wide and begin the arduous process of explaining the project to healthcare providers and signing assistance agreements. This last item included rules for administering and reporting for pharmaceutical supplies that would be essential for us to verify the paper trail to the end beneficiary.
- Through the Croatian Ministry of Health and UN health coordinating network that included the Red Cross, WHO, Medecins Sans Frontier (MSF; in the US referred to as Doctors Without Borders), and other major relief agencies, assess the most acute public health issues among refugees and displaced persons.
- Identify vendors and initiate a purchase for pharmaceuticals; this necessitated knowing several critical requirements for this procurement:
  - facilities, addresses, and points of contact;
  - types and quantities of pharmaceuticals needed;
  - a waybill system for managing distribution and delivery once the shipment arrived.

This all proved to be much more difficult than anticipated, evident in the troubles we faced with that first shipment. The further breakdown of shipments into smaller deliveries proved much more challenging than anticipated due to poorly marked and packaged pharmaceuticals. Except for pallets, the warehouse lacked the appropriate infrastructure for sorting and repackaging the supplies. Our fleet of Suzuki Samurais proved to be inadequate for the task of delivery, even for the initial small purchase, forcing multiple journeys by our staff, at times through unfamiliar and hazardous roads, to deliver supplies. Uncovering these initial headaches helped to ease the pain with later procurements and smooth our delivery mechanism, allowing us to better focus our limited resources on assessing needs and tracking usage.

While taxing, one effort that proved to be a major success was touring the country for the assessment. Visiting hospitals, health clinics, and refugee medical centers to meet with health professionals, doctors, nurses, dentists, and others to discuss their medical needs helped establish an understanding of the project requirements, and ultimately to develop working relationships with them. A chief outcome of this tour was developing a better situational awareness of the country and a deeper understanding of stakeholder needs and circumstances. This initial needs assessment laid the groundwork for a solid project.

One lesson I learned from those days overseas and the years since then that I carry with me in my project management career is that negotiation is an essential skill for a project manager. Project managers are typically engaging in some type of negotiation at many points in the project lifecycle, if not during the course of a working day. Negotiations take place regularly over resources (staff, equipment, materials, space and systems, IT and communications), schedule deadlines, and the ever-present scope of work and managing—essentially negotiating the project—through inevitable changes. In those early days I would be negotiating to get computer shipments expedited (we only had one 386 laptop when I arrived, mine,[8] which was shared among seven people in my office), for office space, for warehouse space, with client hospitals and health clinics, with our operating local partner, and with our accountants, to name just a few. These were made more complicated by the ongoing civil war, cultural and language barriers, and, quite honestly, my youth.

I was barely 24 years old and managing a large-scale emergency response project at a national level far from home, with little support and guidance, in a country that was not familiar to me (take note International Relations majors; a degree does not confer cultural literacy nor competency). What I lacked in experience, I made up for in perseverance and patience. Listening and focusing in on the needs of your negotiating partner is one of the key ingredients to a successful negotiation, and during those early assessment trips into the field, I did a lot of listening and note-taking, meeting not only with health professionals, but also with leaders in the community and with the community members to hear their stories and understand their circumstances. Once you appreciate where your counterpart is coming from and what you and your partner want to achieve, then it is a matter of finding common ground on which you can both build a mutual understanding through which you can then both accomplish your goals.

However, negotiation is a skill that is a means to balance expectations between competing stakeholders. It is in stakeholder management where gauging stakeholder importance through influences interest, and their involvement in the project helps to grasp the nuances required to balance these needs, be they information, resources, or outcomes. These stakeholders in rank order were as follows:

1. **The funding agency:** The US State Department Bureau for Refugee Programs; without them, funding would not have been available, and there would not have been a project.
2. **The end-recipient (beneficiary):** Identifying and considering their needs was critical since they were the end beneficiaries we were serving.
3. **Client health facilities (hospitals, clinics, etc.):** They were a critical partner without whom there would be no way to deliver the medical resources we were providing.
4. **Implementing partner (Caritas Croatia and other local humanitarian agencies):** We had to work with a local partner for HR management, and without their cooperation

and support, we would not have had the facilities or project staff to implement the project.[9]

5. **Project team**: Without the team and their commitment, the project would not have been implemented.
6. **Croatian Government**: Without their authorization to operate in the country, we could not have been operating.
7. **Vendors**: Without them, we would have no supplies and means, warehouses, trucks, etc. to deliver our relief assistance.

This may appear to be obvious, but managing these competing interests was no simple task.

GETTING RELIGION AS A PM – GETTING MY PMP®[3]

Years later when I was in business school, I had the opportunity to study project management as an elective, part of my Master's Degree in the Science of Management.[10] My good fortune was having one of the acclaimed practitioners of project management, Prof. Hans Thamhain,[11] teaching the course. Prof. Thamhain started with the basics of project management and slowly built on those concepts. As he laid out the project management methodology, the experience I had gone through those previous five years started to click into place and make collective sense to me as part of this framework. Scope and stakeholder management, team development, cost estimating techniques and forecasting, and project quality and evaluation methods and tools became starkly evident as part of a larger, extensive, and cohesive framework.

Leveraging those lessons while working as a management consultant a few years later, I used those methods and tools of project management to great effect in my practice. Eventually, I decided to get my PMP® certification from the Project Management Institute®. I prepared for it like anything else, making it a project. Setting the goal was easy: passing the exam and getting my PMP. Of course, there was the application process, which included a listing of all major projects undertaken during your years of experience. Then it was studying, assessing my knowledge, and clocking my time on practice exams, benchmarking my progress along the way to ensure I would walk in and pass the exam, which in 2003 I managed to do.

While some people may find it corny, I keep a copy of the PMBOK® on my shelf close by and hang my PMP® certifications proudly on my wall and, more importantly, take the discipline of project management seriously. As one of my colleagues in project management has said "Project management is a common sense approach to projects, but it requires uncommon discipline" to be effective.[12] What I have learned over my 30+ years as a project manager is that, while seemingly simple, the project management methodology— what is considered to be the better practices as espoused in the PMBOK®—is a powerful approach to getting projects done effectively. I have seen the good, the bad, and the ugly in projects; that latter most often rears up when the simple methods and tools are ignored or paid short shrift, leading to the expected project disaster.

Managing emergencies and emergency plans is no different. The only difference in managing those types of responses to events and project endeavors is that they have higher stakes, with life and property on the line if a project is mismanaged. It is with this intent that I write this book, as an advisory and, if helpful, a set of guidelines and examples that may help others who are tasked with planning for emergencies (for prevention, mitigation, preparedness, and recovery) and who need to respond to an emergency and think about near term response and long-term plans for recovery.

NOTES

1 The organization was Catholic Relief Services (CRS), a major global humanitarian agency.
2 All names used in this book have been changed to hide identities except where specifically noted or attributed.
3 The project was funded through the US State Department's Bureau for Refugee Programs which would be later renamed the Bureau for Population, Refugees, and Migrants.
4 The term refugee applies to a person who has fled across an international border, from one country to another, while the term displaced person is used for a person who is internally displaced within their own country. Most, if not all, of the refugees who fled into Croatia came from Bosnia-Herzegovina, while the displaced persons came from within Croatia, mostly from areas within or close to the UN Protected Zones (so-called Krajina and Eastern Slavonia).
5 The Croatian coast has been and is once again today a popular tourist destination and filled with hotels that line the rocky banks of the Adriatic Sea.
6 The Project Management Body of Knowledge (PMBOK®) is a recognized set of best practices and published every four years by the Project Management Institute to reflect the current state of practice among leading project management professionals globally. See www.pmi.org for more information.
7 The Suzuki Samurai was not sold in the United States at the time since it was deemed unsafe as it was prone to tipping over; we dubbed them a "tin can with an engine" since they were such a flimsy four-wheel drive vehicle.
8 The one which was assigned to me by my organization.
9 In fact, we were successful in getting the project funded twice more in the years that followed, somewhat unprecedented for what was considered a one-time grant.
10 The MSM at the Arthur D. Little School of Management is now the MBA at the Hult International School of Business located in Cambridge, MA.
11 Prof. Hans J. Thamhain, October 1, 1936 to July 11, 2014 (age 77). He tragically died in a bicycle accident. A memorial page can be found at: www.morrissouthboroughfuneral.com/obituary/Hans-Thamhain
12 Attributed to Stephen Gershenson, a former instructor at the American Management Association.

Chapter 2

Why Project Management Is a Good Fit for Emergency Management

A good majority of work time for most emergency management professionals or managers in public safety, except for those engaged in first response (frontline fire, police, EMS), is engaged in planning and preparation, and a very small percentage of time and effort in response. Those efforts, whether in prevention, mitigation, or preparation, are most often undertaken through projects; projects planned and funded through tax levy dollars or government grant funded programs, and occasionally through public/private initiatives. Let's explore project management in greater depth.

WHAT IS A PROJECT AND PROJECT MANAGEMENT?

While projects and project management have been defined many times in all sorts of publications, they are both more definitively and simply defined in the Project Management Body of Knowledge® (PMBOK). A project as defined in the PMBOK® is "a temporary endeavor undertaken to create a unique product, service, or result".[1] The definition goes on to further elaborate what each of the parameters means, e.g. temporary and unique. In short, a project has a beginning and an end and is designed to create something that has not been done before or is different in some way than what came before. The definition of a project stresses that it is not routine as you might find in operational work, for example manufacturing a product or the delivery of a service on a regular basis. Projects require substantial thinking to tackle the problems that they are designed to solve.

I would further define a project as an undertaking that requires a degree of coordination among a group of people, usually more than three, across functional disciplines (e.g. marketing, finance, and operations). A project is also typically an undertaking that requires more planning and work than a couple of weeks, although I will explain further in this book that an emergency response can be managed as a type of project albeit with a suitable adaptation of the methodology. Indeed, larger emergency responses usually span more than a couple of days, usually weeks after the disaster event has taken place, when it merges into the recovery phase that may take weeks, months, if not years to come to a conclusion.

This leads to the next important definition, and that is *What is project management?* Again, going to the PMBOK®, project management is defined as "the application of

DOI: 10.1201/9781003201557-3

knowledge, skills, tools, and techniques to project activities to meet project requirements".[2] Project management at its core is about managing expectations; managing expectations, first and foremost, of the client, and secondarily, managing the expectations of other key stakeholders: sponsors (who may be clients or internal management), end-users, team members, vendors, etc. At times, those expectations may come into conflict with one another and that requires negotiating between those expectations to bridge the conflicts, from initiating the project to the time it is closed. A skilled project manager is constantly holding those expectations in line while balancing the budget, keeping the project within scope and on time. This is no small feat in normal operating environments, and considerably more challenging in the world of emergency preparedness and response.

DISASTERS, CATASTROPHES, EMERGENCIES, AND EMERGENCY MANAGEMENT

On the heels of defining terms for projects and project management, at this point in the book it is important to define what is meant by emergency management and what prompts the need for emergency mitigation measures and responses. Much like routine or process-based work, there are routine problems or issues that arise in the course of daily work. Typically, these disruptions to daily operations are addressed by individuals or a small team, managed using resources readily available, and are able to restore systems back to their normal state in short order. Examples of these types of incidents are a temporary power outage, minor flooding, or limited fire outbreak.

A disaster, as defined in the Merriam-Webster Dictionary, is "a sudden calamitous event bringing great damage, loss, or destruction".[3] Unlike minor disruptions, a disaster, depending on the severity, usually requires substantial resources (personnel, material, and equipment) that are not normally readily available to be marshaled and deployed to address critical needs in order to restore a system (operations, businesses, etc.) back to its normal order. The effort required by these resources often extends to days, weeks, months, and perhaps even years. More importantly, the distinguishing feature of disasters is that they require a significant level of coordination as they tend to bring together cross-functional and diverse groups of resources, from different organizations, government agencies, volunteers (NGOs, PVOs, CBOs, and FBOs[4]), government agencies, and companies to work together.

Now let's look at what constitutes a catastrophic disaster. While it may seem like an act of semantic interpretation, in the world of emergency management catastrophic disaster events are quite different. A catastrophic disaster is typically one in which the impacts of the disaster exceed the capabilities of the jurisdiction to respond via the collective efforts of all emergency response and recovery agencies that are part of that jurisdictions' emergency plan and/or operations. What also differentiates a catastrophic disaster event is the degree and period for recovery, which is usually extensive. While routine (or lower level) disaster events can be handled locally or within their normal routine, catastrophic damage can take many months, years, and in some cases decades to recover from and includes a wide area of impact and a significant outlay and funds from both public budgets and private insurers and savings. In some cases previous inhabitants never return and jurisdictions may never fully recover.

A good example of a catastrophic disaster would be the EF5[5] tornado that struck Joplin, Missouri on May 22, 2011; the tornado stretched close to a mile wide and ran a path over 21 miles long and stayed on the ground for close to 20 minutes, destroying over 7,000 homes and numerous municipal buildings, schools, and the main hospital,[6] and killing 160 people.[7] This would be a catastrophe by any stretch of the imagination, but what distinguishes this more than just that was the help needed outside the jurisdiction of the City of Joplin.[8] This help came from neighboring jurisdictions, State and Federal agencies, and major volunteer organizations. Three years following the event, nearly 90% of homes were rebuilt, demonstrating the long period of recovery from such a catastrophe.

An organized response to the critical needs of life-safety, protection and restoration of property, and mitigation of disaster impacts as a result of any major or even catastrophic disaster is carried out through a system of emergency management. Emergency management with its constituent components of planning, prevention, mitigation, preparedness, and response is focused on the coordination of what are called emergency response functions. These emergency functions are broken down into 15 different areas according to FEMA,[9] and these functions are related to life-safety response or, simply stated, preserving human lives and public health: security (law enforcement) and safety (fire service and emergency medical services [EMS]). As referenced above, the secondary missions of any emergency response is to restore property and to mitigate or eliminate the impacts as a result of a disaster. We will explore these further on in this book.

In the sphere of emergency management, this coordinated effort is termed a unity of effort, where these diverse groups work in a coordinated effort toward a common goal. As mentioned in the last chapter, the common working methodology for managing emergencies is called the Incident Command System (ICS), designed to provide a common approach to integrate and coordinate multiple and various resources toward a mutual effort; this system allows the incident command or command element, i.e. the person or people in-charge, to not only effectively direct an emergency response but also to scale up or down the response depending on the situation. While this book is not intended to specifically cover the methodology and terminology of the incident command structure, we will address some of these concepts and techniques as necessary later on in this book in Chapter 15.

The degree to which coordination and management structure necessitates a robust management structure is largely dictated by the size and complexity of the disaster. As outlined in Table 2.1, the National Incident Management System (NIMS) defines the complexity of an incident and the related emergency response along the following parameters.

FEMA defines Incident Complexity Incident and/or event complexity to determine emergency and incident response personnel responsibilities as well as the recommended audience for NIMS curriculum coursework delivery.[10] It is also important to take this incident typology into consideration for emergency planning and preparedness activities, in particular as they relate to the need for project management.

Table 2.1 NIMS Incident Typology

Incident Type	Incident Description
Type 1	• This type of incident is the most complex, requiring national resources for safe and effective management and operation. • All command and general staff positions are filled. • Operations personnel often exceed 500 per operational period and total personnel will usually exceed 1,000. • Branches need to be established. • A written incident action plan (IAP) is required for each operational period. • The agency administrator will have briefings and ensure that the complexity analysis and delegation of authority are updated. • Use of resource advisors at the incident base is recommended. • There is a high impact on the local jurisdiction, requiring additional staff for office administrative and support functions.
Type 2	• This type of incident extends beyond the capabilities for local control and is expected to go into multiple operational periods. A Type 2 incident may require the response of resources out of area, including regional and/or national resources, to effectively manage the operations, command, and general staffing. • Most or all of the command and general staff positions are filled. • A written IAP is required for each operational period. • Many of the functional units are needed and staffed. • Operations personnel normally do not exceed 200 per operational period and total incident personnel do not exceed 500 (guidelines only). • The agency administrator is responsible for the incident complexity analysis, agency administration briefings, and the written delegation of authority.
Type 3	• When incident needs exceed capabilities, the appropriate ICS positions should be added to match the complexity of the incident. • Some or all of the command and general staff positions may be activated, as well as division/group supervisor and/or unit leader level positions. • A Type 3 IMT or incident command organization manages initial action incidents with a significant number of resources, an extended attack incident until containment/control is achieved, or an expanding incident until transition to a Type 1 or 2 IMT. • The incident may extend into multiple operational periods. • A written IAP may be required for each operational period.
Type 4	• Command staff and general staff functions are activated only if needed. • Several resources are required to mitigate the incident, including a task force or strike team. • The incident is usually limited to one operational period in the control phase. • The agency administrator may have briefings and ensure the complexity analysis and delegation of authority are updated. • No written IAP is required but a documented operational briefing will be completed for all incoming resources. • The role of the agency administrator includes operational plans including objectives and priorities.
Type 5	• The incident can be handled with one or two single resources with up to six personnel. • Command and general staff positions (other than the incident commander) are not activated. • No written IAP is required. • The incident is contained within the first operational period and often within an hour to a few hours after resources arrive on scene. • Examples include a vehicle fire, an injured person, or a police traffic stop.

PROJECT-BASED WORK AT MANAGING RISK – MITIGATION AND PREPAREDNESS PROGRAMS

Another way of looking at projects, especially in terms of emergency management, is in terms of risk. In brief, a risk on a project is an event that can have a positive or negative impact on the project objectives. Risk in projects is a factor of the investment of time and resources (typically monetized in some way) to yield a positive result. So in a sense, the

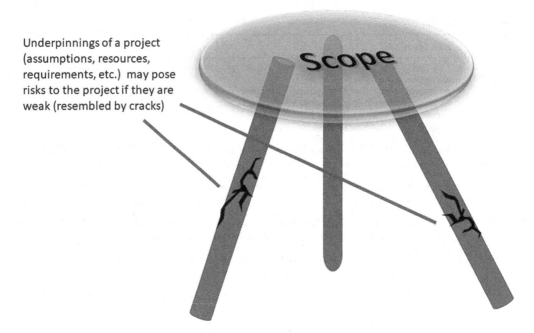

Underpinnings of a project (assumptions, resources, requirements, etc.) may pose risks to the project if they are weak (resembled by cracks)

Figure 2.1 Legs supporting a project scope.

organization (whether government, non-profit, or private) is *risking* that investment for what may be deemed a *positive risk* or the benefit gained; the positive outcomes of risk are not always recognized as such although they are certainly present. Unfortunately, risk management is one of the undervalued and under-practiced areas of project management, partly, because many assumptions are taken for granted or important factors are ignored, only to transform into negative risks when the underlying factors that underpin them change, as Figure 2.1 illustrates.

A good example from my own experience was in 1994 when my project was in sore need of an epidemiologist to assess the health status of our client population, and good fortune (a positive risk) brought an epidemiologist from Sarajevo to our doorstep. I had happened to meet her there while I was conducting a short mission to evaluate the needs for an office. She was soon working for our project and solved the niche we needed to fill. I had assumed she would stay with the project for the foreseeable future (that being a year or two). This assumption turned out to be a risk when, after a year in our employ, she informed me that she would be emigrating to Switzerland. We will delve further into risk management in Chapter 3.

While the definition of risk may appear clear to most readers, I have found in my years of consulting and teaching on project-based work that there is some confusion. Again, going to the PMBOK®, a risk can be simply defined as an event that can negatively or positively impact the project in terms of schedule, cost, or scope (the triple constraint). Usually, a risk that primarily impacts one of the constraints winds up affecting the others. Many confuse concerns with risks, such as "the team may not communicate effectively"; this would generally be considered to be a project management quality issue (*note:* not a product quality issue specifically). If, however, the ineffectiveness of team communication results in the wrong deliverable being sent to the client, e.g. an incorrect design document, then that act (wrong item sent) may be defined as a risk event with associated negative consequences.

Risks in the world of project management are made up of two major components: probability, the likelihood that a risk may occur, and impact on the project, stated initially in monetary terms[11] although there are also schedule impacts (which typically result in monetary impacts). It is through this lens then that risks are identified, defined, prioritized, and managed throughout the lifecycle of the project.

Naturally, the world of emergency management is primarily focused on managing critical risks whether to public safety, health, overall well-being, economy, etc., and proper functioning and the multiple systems that support those ends.

EMERGENCY RESPONSE IS A PROJECT, BEGINNING TO END

All emergency situations, from a low-level disaster to a catastrophic event, have a lifecycle: a beginning and an end. It starts with the trigger of the disaster incident, at times a slow start, such as a storm to those which are sudden, like a tornado, and proceeds through the stages of impact through to recovery. Whether we are looking at those with notice, such as hurricanes, or those with no notice, as in earthquakes, there exists a state of readiness, a response, and recovery to return life to its normal state or what might be called a new normal. In most cases, major or catastrophic disasters leave their mark and change the way of life in the communities they impact.

Much like with a wedding, you can spend months (in some cases years) planning and preparing for the big day or, like some spontaneous couples, get married by a justice of the peace and go to city hall for the marriage license with little ceremony and fanfare. Any efforts prior to or after the event become part of the work. Building levees, instituting earthquake safety standards, creating flood maps, these are all measures taken to mitigate against the damage that natural disasters may cause. Planning, which is the focus of much of emergency management in preparedness, is a complementary component to mitigation and, in some cases, may be looked at as an extension of the mitigation and prevention phase. Once mitigation, prevention, and preparedness measures have been put in place, then any disaster event that arises that requires those emergency responses is put into action. Table 2.2 presents the emergency management process and related projects and examples.

THE ROLE OF THE EMERGENCY MANAGER

Unlike projects in the private sector that are oriented to achieving a return on investment, many projects in the government domain, as well as companies, are aimed at achieving benefits for the public good, clients, customers, or stakeholders which are sometimes difficult to quantify in terms of a return on investment. Emergency mitigation or preparedness projects have been termed "Spending money that governments (and even companies) claim they don't have on things that they do not want to happen". The flip side of this is when disasters do strike, then that money and effort is always thought of as having been

Table 2.2 Emergency Management Process and Related Projects and Examples

Phase of EM	Emergency Management Activity or Project; Examples of Project Types
Mitigation and prevention	• Building levies to mitigate river flooding • Relocate communities outside of flood zones • Educating the public to increase awareness of household preparedness
Preparedness	• Developing Emergency Operations Plans • Training staff on emergency plans and protocols • Conducting drills and exercises
Response	• Coordinating the evacuation of citizens from flood zones • Deploying search and rescue teams after a devastating earthquake • Building a berm in advance of river flooding
Recovery	• Debris removal after a tornado • Restoring power after a major power outage • Rebuilding a community after a natural disaster

well spent or is assumed to have been present, when in fact it took years of work, justification, and planning and exercises to pull off a successful response.

In this paradox that elected leaders and government officials find themselves, they are managing the tension as stewards of taxpaying dollars (or other currencies outside the United States) and balancing this against preparing for the response to the exigencies of disasters. Government leaders are accountable to the budgeting process, fiscal reporting, interest and "watchdog" groups, the press, and finally government audit. Most support for long-term emergency mitigation and preparedness programs is funded at the Federal level through grants to the states which distributes this funding and manages it down to the local level, either on a county or municipal level. This cycle takes years, often with grant funded projects and programs being carried out over the same time frame, at times with seemingly unrealistic deadlines when approvals run late and/or get mired in legal contracting between emergency agencies and vendors.

Many emergency managers have risen through the ranks, from the field level, as an EMT, firefighter, or police officer through the ranks to a captain, chief, or senior leader in their organization. Most of the work that these middle and line managers in first response (law enforcement, the fire service, and emergency medical service) is operational response: responding to an emergency call, debriefing and reporting on these calls, training to improve their capabilities, and daily administration, maintenance, and support for these operations.

New projects, such as creating new public safety programs, adopting new methods, techniques, or technologies, developing an exercise, require aspects of project management: managing project scope, timelines, and budget, which may not be familiar skills for many of these managers and leaders, so the guidance in this book is intended to add to their toolkit. As many colleagues in the field have related to me, the greater percentage of work a professional emergency manager is engaged in on a regular basis is project based, and a much smaller percentage in response (with the exception of the pandemic).

Additionally, in major disaster responses, when managing different functional responses, emergency managers, whether operating within a Unified Command or Single Command in cooperation, are working cross-functionally, marshaling limited resources toward a common goal within operational periods and interim objectives, but with an unclear initial scope and undefined end date and exit strategy; all aspects of project management, albeit best described as "rolling wave project management".

In the United States, emergency management in response to larger events, with multiple operational periods and multi-agency response, has largely been governed by the Incident Command System (ICS) since its adoption as a methodology in the 1970s and development in the ensuing decades, incorporated as part of the National Incident Management Systems (NIMS).[12] While ICS does address the needs of managing multi-agency, public, and private organizational coordination for large complex emergency events, there is a need to fill gaps when it comes to understanding and managing the scope of new projects (training, adopting new technology such as ICS software, or the larger management of drills and exercises); and managing human resources, budgets, schedule, procurement, quality, communication, and risks related to project management (not the usual operational risks). The full spectrum of project management methods addresses these needs through an integrated approach to these managerial activities.

A couple of items I want to note before the moving on to rest of the book. While this book is intended to explain the connection between project and emergency management and provide a solid context, it is not intended to cover these two subject areas, or the related areas of risk management or leadership, in exhaustive depth. There are numerous books and standards that delve into these areas, some of which I reference in this book. Although most of the references, frameworks, and methodologies are based on US models, one of my objectives of the book is to target a global audience. I have referenced frameworks used in other parts of the world, ISO, UN, etc., so no matter where you may be based on the globe, professionals in emergency management and public safety will be able to understand and apply these concepts and methods in their own context. Lastly, some concepts may appear to be only relevant to one audience, so I may at some points in this book answer the question as to "why this is important", explaining the relationship of a concept or method to the current state of common practice.

NOTES

1 *Project Management Body of Knowledge, 3rd Edition*, 2004 Project Management Institute, Four Campus Boulevard, Newtown Square, PA, p.5.

2 *Project Management Body of Knowledge, 3rd Edition*, 2004 Project Management Institute, Four Campus Boulevard, Newtown Square, PA, p.8.

3 From Merriam-Webster: www.merriam-webster.com/dictionary/disaster?utm_campaign= sd&utm_medium=serp&utm_source=jsonld

4 NGO=Non-Governmental Organization, PVO=Private Voluntary Organization, CBO= Community-Based Organization, FBO=Faith-Based Organization. Each of these terms describes a type of organization that comprises both paid and volunteer staff; NGO is typically used internationally and describes humanitarian relief, response, and development organizations with paid staff.

5 EF stands for Enhanced F Scale for Tornado damage. This rating scale is similar to the Saffir-Simpson scale for hurricanes in that the skills are based on ranges of wind gust estimates based on an evaluation of damage, usually taken immediately after a tornado event. The F scale was originally developed by T. Theodore Fujita of the University of Chicago in 1971. This was subsequently updated in 2007. Reference: www.spc.noaa.gov/faq/tornado/ef-sclae.html

6 NCDC Event Record. NCDC Storm Events Database. National Oceanic and Atmosphere Administration, National Climatic Data Center. See www.ncdc.noaa.gov/stormevents/eventdetails.jsp?id=296617

7 McCune, Greg (November 12, 2011). "Joplin tornado death toll revised down to 161". Reuters. See www.reuters.com/article/us-tornado-joplin/joplin-tornado-death-toll-revised-down-to-161-idUSTRE7AB0J820111112

8 Powerful tornadoes kill at least 31 in US Midwest. Kevin Murphy (Reuters) May 22, 2011. See www.reuters.com/article/us-usa-weather-tornadoes/tornado-devastates-joplin-missouri-116-dead-idUSTRE74M08L20110523

9 www.fema.gov/media-library/assets/documents/25512

10 National Incident Management System Incident Complexity Guide: Planning, Preparedness and Training (fema.gov). See www.fema.gov/sites/default/files/documents/nims-incident-complexity-guide.pdf

11 As we will get into later in Chapter 4, the monetary term used is the expected monetary value (EMV).

12 Gil Jamieson (2005) NIMS and the Incident Command System. International Oil Spill Conference Proceedings: May 2005, Vol. 2005.

Chapter 3

Risk Management for Emergency Management and Public Safety

"We are shown that disasters are not natural. Only the threat is natural, so we have to manage the risk. We have to manage our vulnerability. We have to manage the exposure of people, and we have to manage how we use the land; how the earthquakes and the tsunami and the volcanoes can impact us. And how people interact with that natural environment. We take decisions every day, and most of these decisions expose people to danger, especially vulnerable people. ... It's not only a matter for an 'Office of Emergency Management'. It is a matter of the Housing Ministry, the Development Ministry, the Economy. It's about poverty. It's about culture; and about education; (and) is about immigration. There's a lot of issues that produce the disaster."

Interview with Victor Orellana Acuña, Disaster Management Consultant and former National Deputy Director at the National Emergency Management Office for Chile; 2019

Threats and hazards, and the disasters they cause, are the basis of the emergency management profession. However, not all hazards pose a threat and result in disasters, and it is our understanding of the dynamics behind each type of hazard that allows us to analyze their potential impacts, our vulnerability to those impacts (exposure) that emanate from the hazard, and then reduce our exposure by ultimately managing the risk it poses.

Threat and hazard identification and risk assessment (THIRA) is the foundational analysis of modern risk management. When a hazard threatens or otherwise reacts with the human built or natural environments, the risks associated with that hazard can be measured and assessed. Understanding the risk posed by identified hazards allows emergency managers to take action where it is needed, whether to prepare for or mitigate against a future occurrence.

When a hazard is realized, the risks from that hazard can result in an emergency situation, or worse, a disaster, a non-routine emergency event. This in turn will necessitate a larger scale emergency response, usually over an extended period of time (what are called operational periods in emergency response[1]), and recovery efforts to ensue soon after, until life is returned back to normal (or a new normal). Once an after-action review is performed, the new information gleaned from an actual event will subsequently improve the risk analysis process and allow for adjustment in risk treatment and preparedness programs.

Before we go into greater depth on the topic of risk management (including threat and hazard analysis), I want to be clear that I am covering this topic at a sufficient level here and in Chapter 4 to set the stage for Chapter 5 on scope development. In other words, to

DOI: 10.1201/9781003201557-4

identify the major disasters, impacts, and capabilities needed to start addressing them with sensible, justifiable, and practical plans to enhance these capabilities. This is not intended to exhaustively cover all of the potential risk management methodologies currently used for these areas, such as Comprehensive Preparedness Guide (CPG) 201: THIRA,[2] the business-related practice of risk assessment and business impact analysis (BIA) embodied in such business continuity standards such as ISO 22301 Societal security –Business continuity management systems – Requirements,[3] or National Fire Protection Association (NFPA) 1600 Standard on Continuity, Emergency, and Crisis Management.[4] These methodologies are publicly available, targeted to a specific audience and set of use cases, and are elaborated well enough to cover all of the essential methods and tasks. There are plenty of books written on the topic as well. For the practitioner who wants to immerse themselves in this area, I would recommend reading these standards and books to gain a broader and deeper perspective.

One other point I want to make clear before we delve into the area of risk management. This chapter addresses risk management from the perspective of making the business case for the types of projects that might be considered for investment by a jurisdiction (government, agencies, etc.) or an organization. In Chapter 7, we will address risk management from a project perspective, a much narrower focus. Although the methodologies and tools are similar, the level and complexity are quite different.

WHAT IS A RISK?

Risk is poorly understood, and the term is often used incorrectly, so let's step back a moment and go over some basic definitions, first by defining the term *risk*. A risk is an event which can have a negative or positive impact if it occurs. It is often defined by two major parameters:

- Probability, also called likelihood, that a risk may occur; and
- Impact, also called consequences; this may also be termed the potential exposure to the risk, in other words the loss that may be experienced if the risk occurs.

Let's pause here and differentiate and further elaborate the definition of risk. Notice that I used the phrase *risk is an event*. It is important to draw the distinction between a risk event and what constitutes a risk broadly. In Table 3.1 the Bowtie Analysis illustrates an analysis of a flood event that helps to support this point. It helps to puts into context what emergency managers sometimes refer to as what is "Left of Boom" and "Right of Boom"; this is a military term used to describe the preparation before the "explosion" or incident starts, and the impacts from the aftermath, the response, and recovery.

The simplified Bowtie Analysis chart given below demonstrates a linguistic device that I teach in my risk management courses: *the risk metalanguage.*[5] The risk metalanguage is used to parse the descriptive language to better define risk events and consists of the following basic construct:

Because of <one or more causes>, <risk> might occur, which would lead to <one or more effects>.

Table 3.1 Bowtie Analysis

Causal Factor(s)	Event	Direct Impact (Consequences)
Heavy rain creates riverine flooding which inundates the town flood plain	*Houses in the flood plain flooded with 3 ft. of water*	Foundation and structural damage to homes; electrical outages; property damage (furniture/ appliances) mold growth Drowning Infection from polluted water

In this simple case, because heavy rain (causal factor) creates riverine flooding in the flood plain, houses in the flood plain become flooded with 3 ft. of water (risk event), which would lead to damage to the foundations, structures, electrical outlets, other property, and potentially allow mold to grow (one or more effects). Of course, analyzing this risk is more complicated than this basic analysis, and we need to take into account other considerations, such as vulnerabilities, risk responses in place, including mitigation measures and risk transfer, which we will get to later in this chapter.

I use the risk metalanguage to describe risks as I often hear causal factors, or even consequences, described as risks themselves, which muddles the very reason why we define and analyze risks in the first place. For example, in all types of settings, I have often heard the following wording used to describe risks (note – where it falls into the risk metalanguage is given in parentheses, and an example):

- His arrogance and poor project management (causal factor; example: the Panama Canal Project managed by Ferdinand De Lesseps[6])
- Not enough technical knowledge (causal factor; example: use of drones)
- Flood barrier construction delays (consequence for a project [not a hazard risk])
- Damaged utility poles (consequence; note – damaged utility poles may become a causal factor for other risks, such as a power outage or fire)

WHAT IS RISK MANAGEMENT[7] AND WHY DO WE MANAGE RISKS?

We manage risks in any area of work in order to address consequences of risk events, either to avoid the pain (loss, damage, etc.) or enhance the upside (of opportunities, also called positive risks). We manage risk events in order to improve our decision-making in order to prevent (avoid), mitigate, and/or transfer risks or accept the risks and deal with them when they happen (rarely a good idea, but it largely depends on the risk event). Decision-making in the context of business, governments, and any organization is how we will use and invest our resources, money (funding) which is in direct relationship to how we allocate people, systems, equipment, materials, services over time.

We typically look at risks through the lens of what are called pure risk, risks that are only associated with expected negative consequences (or loss) of what might happen if a risk event occurs. *Pure risks* have no upside to them. *Speculative risks* have both positive and negative aspects to them; for example in the financial world, making an investment has both positive and negative returns, an upside or downside. While positive risk aspects do exist, often called opportunities, in the world of emergency management and public safety, we consider only the pure risks when we are assessing risk, with only the downsides to consider; in short, we allow other people in other elements of society to speculate on and manage the upsides of risks.

BEYOND PROBABILITY AND IMPACT

In addition to probability and impact, there are other factors to consider when it comes to risk events related to disasters and public safety issues. The following factors are also key to understanding the dynamics of a risk event as it will often dictate what type and extent of resources and planning we must engage in to address it:

- **Velocity:** speed of onset of the risk
- **Time horizon:** duration that the risk may last
- **Correlation:** relationship between risks and risk factors
- **Geographic range:** how geographically widespread the risk is
- **Volatility:** to what degree is the risk event subject to change

Velocity

Velocity is the speed of onset of the risk. There are certain risk events, threats, and natural hazards, for which we have some degree of time before they occur, and commensurate warning through weather forecasts, such as hurricanes, riverine flooding, and extreme heat. Then there are other threats which provide little to no warning, such as tornados (although forecasts and warnings are improving), wildfires, volcanic eruptions, and terrorist attacks. The velocity factor is important in our consideration of how we prepare for such risk events. It is also important when considering the triggers for emergency responses and the decisions that deploy resources into emergency operational responses.

Time Horizon

The time horizon is the duration that the risk event may last. A shorter time horizon event may be a tornado, earthquake, or active shooter situation where it starts and stops in a short period, from seconds to minutes in a given location. A longer duration risk event might be a hurricane, wildfire, or a pandemic. The time horizon factor is important as it drives the length and the extent of the response that need to be sustained. This does not mean that the impacts from a disaster event may be short term. Impacts may linger far longer than the event itself, at times, days, weeks, months, or years after a major disaster.

Correlation

Correlation is the relationship between risks and risk factors. Risk events do not exist in a vacuum, and when a major hazard poses a threat, there are multiple risks they pose, some of which may interact or correlate to one another. For example, a hurricane produces high winds, heavy rains, and, depending on the location and timing, flooding. Each of these three main threats poses its own risks. High winds are clearly correlated to coastal flooding in a hurricane, as the wind whips the sea onto shore. Correlation is an important factor to consider when looking at how we will manage risks through standard options, mitigation, transfer, even avoidance where applicable. Correlation is also a helpful factor for emergency managers to consider for various types of emergency events, in particular looking at the bigger picture and tertiary effects. For example, in the pandemic we are seeing a correlation between quarantine measures and psychological illness.

Geographic Range

Geographic range is defined by the impacted area that an emergency event can reach. For example, Hurricane Sandy (see Figure 3.1) stretched from the Mid-Atlantic to Cape Cod and as far inland as the Appalachian Mountains, causing snow to fall in that region. In contrast, the 2011 Joplin Tornado, one of the deadliest tornados to strike in the United States and rated an EF5, lasted 38 minutes and tore a path 22.1 miles long and between ¾ to a mile wide, destroying a good part of downtown Joplin, MO, killing 158 people and

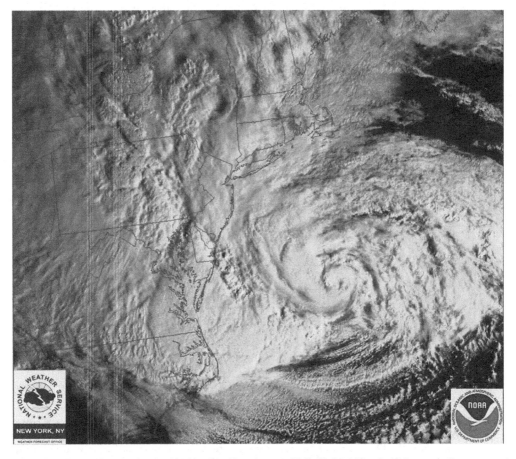

Figure 3.1 Super Storm Sandy Visible Satellite Image 2012-10-29. (Credit: National Oceanic and Atmospheric Administration/National Weather Service.)

injuring over a 1,000.[8] This does not mean that the impacts from a disaster event cannot be felt far from the center of where it occurs. In fact, massive wildfires in the Western United States have created polluted air on the Eastern Coast of the United States in recent times.

Volatility

Volatility is the degree to which the risk event is subjected to change. When we are looking at different natural hazards, we usually understand the latitude of change that a hazard will likely have, whether it is a hurricane or a flood, even some wildfires; we expect that some hazardous events will come and go. However, we are seeing increasingly complicated disaster events, which are amplified by global warming, a destabilized climate, human encroachment into risk prone areas, and increased reliance on technology and vulnerabilities. When Hurricane Harvey hit the Houston area on August 25th, 2016, the assumption was like other hurricanes it would sweep through in a standard 12–24 h window. Instead, Hurricane Harvey sat over the Houston area for four days producing record rainfall, as high as 5 ft. in some areas, causing devastating flooding and over $125 billion in damage.

The natural hazard that is most identified with volatility, at least until the COVID-19 pandemic, has been wildfire. In fact, wildfire science has been and continues to be studied in-depth. As Figure 3.2, developed by NIST and the National Fire Service as a hazard scale for wildland fires illustrates, there are a number of factors which can determine the velocity and geographic range of a wildfire: from windspeeds, to humidity and temperature levels, to wildland fuels, and the topography over which it may travel.

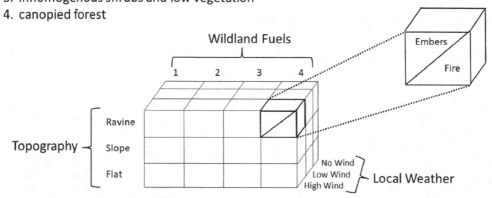

1. homogeneous surface fuels
2. inhomogenous surface fuels
3. inhomogenous shrubs and low vegetation
4. canopied forest

Figure 3.2 Matrix for capturing exposure from wildland fuel. (From NIST and Forest Service Create World's First Hazard Scale for Wildland Fires December 05, 2012. See www.nist.gov/news-events/news/2012/12/nist-and-forest-service-create-worlds-first-hazard-scale-wildland-fires; Credit: Alex Maranghides/NIST.)

HAZARDS AND THREATS

Now let's move on to a discussion of hazards and the potential threats they may cause (risk events) and how they are defined. FEMA[9] defines threats and hazards as follows:

- **Hazard:** Something that is potentially dangerous or harmful, often the root cause of an unwanted outcome. An event or physical condition that has the potential to cause fatalities, injuries, property damage, infrastructure damage, agricultural loss, damage to the environment, interruption of business, or other types of harm or loss.
- **Threat:** Natural, technological, or human-caused occurrence, individual, entity, or action that has or indicates the potential to harm life, information, operations, the environment, and/or property.

The key difference between the two is that hazards emanate from natural causes such as extreme weather, seismic events, or nature, whereas threats emanate from humans, either by accident or intentionally, for example an act of terrorism. In practice, these terms are often used collectively or interchangeably, even though they should not be.

Hazards and threats can be broken down into two major categories: natural and human caused (what was once called *manmade*).

Natural disasters can be:

hurricanes	tsunamis
storms (derechos, Nor'easters)	volcanic eruption
floods	winter storms
earthquakes	extreme cold/extreme heat
tornadoes	drought
wildfires	pandemic
land/mud/snow (avalanche) slides	

and human-caused disasters:

technological hazards or accidents	chemical
structural fires	biological
dam failures	nuclear/radiological accidents/events
hazardous material incidents	terrorism
nuclear accidents	explosive events

THREATS AND HAZARDS: MANAGING RISKS

The natural starting point for any risk management program is scanning the horizon for any potential hazards and threats. The traditional risk management process is given in Figure 3.3.

In standard risk management this process is utilized in business practice. In the governmental or public sphere, this process of what is called "Threat and Hazard Identification and Risk Analysis" (THIRA) is embodied in the CPG 201: Threat and Hazard Identification and Risk Assessment and Stakeholder Preparedness Review (SPR) Guide. This guide is

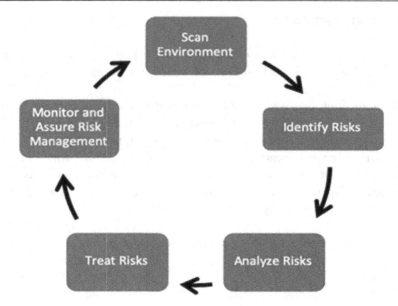

Figure 3.3 The risk management cycle. (From Elliott, Michael W. *Risk Management Principles and Practices*, 2nd Edition (The Institutes 2016), p. 5.16; used with permission from the American Institute for Chartered Property Casualty Underwriters.)

published by FEMA as a standardized method for thoroughly planning for hazards and threats.

THIRA outlines steps that must occur in order as follows:

- Identifying the hazards
- Assessing risk for each hazard identified
- Analyzing hazard risks in relation to each other
- Treating hazard risk according to prioritization

These processes are similar to each other, but we have to consider how we are going to manage them at the micro level. How do we treat risks once we have assessed, prioritized, and analyzed them for potential impacts based on vulnerability; we choose among our options: avoid, transfer (usually through insurance), mitigate, or accept. We will delve further into risk treatment in the next chapter.

IDENTIFYING HAZARDS AND THREATS: CLASSIFYING EMERGENCIES, DISASTERS, AND CATASTROPHES

What defines a major threat or hazard from a minor one? Is it disruption and harm or loss to life, if left unaddressed? One way of looking at hazardous events is provided by one of the pioneers of disaster research in the modern era, Dr. E. L. Quarantelli. He defined these terms as provided in Table 3.2.[10]

Chapter 2 provides a typology of emergency incidents from NIMS as a way to measure levels of emergency response. We can differentiate incidents in terms of planning, resources

Table 3.2 Quarantelli Emergency Levels

Incident Type	Description	Event Examples
Emergencies	Routine emergency response that can be addressed within a jurisdiction's capabilities (resources and timeframe).	Multi-car accident (this would be a Type 5 incident; see chart above)
Disasters	Disaster events exceed the capacity of the jurisdiction in the near term. It impacts unfamiliar and larger groups of people who cannot operate their lives in the same way before the event. It calls for a change in the performance level and standards to meet new and increased needs as a result. There is also less of a separation between the private, public, and non-profit sectors, often increasing interdependencies between different sectors of a society, perhaps not accustomed to working with one another.	Getty Fire of 2019 (this would be a Type 2 incident)
Catastrophes	Catastrophes have some of the same characteristics as disasters but differ in that the severity of impacts are extreme that they completely disrupt the jurisdictions' capabilities to respond and recover within a reasonable amount of time and to a level of performance that meets basic needs; this includes impacts to all emergency response resources and critical infrastructure. There are typically long-term impacts that prevent a full recovery of the population in number or in standards of living.	311 Fukushima Disaster (this would be a Type 1 incident)

required over a significant period and prepared in advance, looking at them first through the lens of "Management by Exception". Management by exception is a concept of how to address managerial work by prioritizing the issues that are exceptions to routine operational work; any non-conforming outcomes or process flaws that arise, also called problems, are then highlighted and dealt with. Any work that proceeds as expected without error or increased effort should warrant limited attention by management.

The first criterion as we stated above is the degree to which the incident can impact life-safety (potential lives lost or physical harm), create major loss of property, and the time to bring the hazard under control and to a conclusion to prevent any further threats to life or property. This is what FEMA in the THIRA process (defined above) describes as putting the hazard/threat into context. This is driven by a number of factors:

- Level of the hazard/threat, for example hurricanes are defined by category according to the Saffir-Simpson Hurricane Wind Scale (Cat. 1–5).[11]
- Time of the year (season) and timing of the event (as in the above during the Summer or late Fall; high or low tide), if it occurs during the day or night.
- Where it occurs.
- Vulnerability to the hazard influenced by the level of mitigation in place. This is much more difficult to define as mitigation measures that are in place may vary, such as implementation of building codes in earthquake prone areas.

We can also apply the lens of "Management by Exception" to the basic emergency response areas, by looking at the following factors and assessing whether a major response might engage one or more of the following services over an extended period of time or require additional effort and coordination beyond the existing capabilities.

First Response Agencies

- Public safety (law enforcement)
- Fire service/dept.
- Emergency medical services

We can also look at the disruptive impacts to basic lifelines or critical infrastructure.

Critical Infrastructure/Lifelines[12]

- Food supply chain
- Housing
- Public works and utilities (energy, power, water)
- Telecommunications
- Healthcare and medical services
- Sanitation
- Transportation
- Financial services

RISK ANALYSIS

Risk analysis is a quantitative method for understanding risk events and defining them more clearly for planning purposes. Risk assessment differs from risk analysis in that the assessment is more qualitative, a first pass review, designed to allow us to prioritize risks and start putting them into context, so we can then understand their specific impacts and focus on the priorities, and examine the vulnerabilities and gaps to mitigate, respond, and recover from disasters.

The obvious first priority is to look at potential for loss of life and impacts to quality of life, and then major property damage: residential, commercial, and government and critical infrastructure. We then want to look at the estimated potential for disruption to first response services and lifelines (core capabilities) specifically looking at what level of service may be reduced or halted, how many hours it may be offline, and whether the quality may be limited to damage in some manner to gauge whether it warrants a more robust disaster response.

There are a number of government (in the United States) and commercial systems that model potential impacts from different types of hazards. Some of the common governmental systems used are listed in Table 3.3.

These hazard assessment models, for example HAZUS MH, HAZUS MH 1.4, ALOHA (chemical threats), and other software programs require training and expertise to configure and interpret the data and models. Some jurisdictions, depending on their size, may either use internal resources with the expertise needed (department focused on risk or disaster risk reduction and resilience) or hire contractors to conduct a thorough and in-depth threat and hazard analysis as part of an overall risk assessment. Risk and hazard-specific specialists can provide more detailed and expert analysis of hazards and their impacts. With any risk models you need to be careful when looking at the base assumptions and

Table 3.3 Hazard Types and Simulation Modeling Systems

Hazard Type	System/Resource Description and Source
Drought	NOAA/Nation Integrated Drought Information System (NDIS) www.drought.gov
Chemical emergencies	EPA/NOAA ALOHA Software hazard modeling program for the CAMEO® software suite, which is used widely to plan for and respond to chemical emergencies. www.epa.gov/cameo/aloha-software
Earthquake	USGS hazard and risk assessment maps www.usgs.gov/natural-hazards/earthquake-hazards/hazard-and-risk-assessment
Earthquakes, floods, tsunamis, and hurricanes	FEMA's HAZUS-MH Program (Hazards United States-Multi Hazard) is an emergency planning and response software package. www.fema.gov/flood-maps/products-tools/hazus
Flooding	FEMA National Flood Hazard Layer www.fema.gov/flood-maps/national-flood-hazard-layer
Storm surge	NOAA Sea, Lake and Overland Surges from Hurricanes (SLOSH) model is a computerized numerical model developed by the National Weather Service (NWS) to estimate storm surge heights resulting from historical, hypothetical, or predicted hurricanes by taking into account the atmospheric pressure, size, forward speed, and track data. www.nhc.noaa.gov/surge/slosh.php National Hurricane Center – National Storm Surge Hazard Maps www.nhc.noaa.gov/nationalsurge/#intro
Tornadoes, hail/wind	NOAA's National Weather Service Storm Prediction Center SVRGIS www.spc.noaa.gov/gis/svrgis/
Tsunami inundation zones	Oregon Tsunami Clearinghouse (covers both States of Oregon and Washington coastal areas) www.oregongeology.org/tsuclearinghouse/
Wildfire	The California Department of Forestry and Fire Protection's Fire and Resource Assessment Program (FRAP) https://frap.fire.ca.gov/ USGS and the US Forest Service, the Fire Danger Forecasting Project www.usgs.gov/ecosystems/lcsp/fire-danger-forecast

factors you may be using, especially looking at historical precedents. As mentioned before, given what we know about climate change, we need to question our expectations.

Earlier in this chapter we started to reference the differences between how governmental agencies and the commercial sector, for-profit, and non-profit organizations analyze threat/hazard-based risks. As we begin to explore how they manage them, we will also note these differences. In addition, there are also some differences in how supranational organizations such as the United Nations which I will touch on this at the end of this chapter.

CORE CAPABILITIES

FEMA, and by default many state, local, tribal, and territorial organizations (SLTT), use the **Core Capabilities**[13] to set targets for capabilities along the spectrum of emergency management mission areas: prevention, protection, mitigation, response, and recovery. These core capabilities were established as part of the National Preparedness Goal starting in 2011 under Presidential Policy Directive (PPD)-8. Along with the five mission areas, there are 32 core capabilities that define the SMART targets for each capability. The THIRA analysis is linked to this and uses the same method. An example from the THIRA document is shared in Figure 3.4.[14]

Example Estimated Impacts for Core Capabilities							
Prevention	Protection	Mitigation	Response			Recovery	
Screening, Search and Detection	Access Control and Identity Verification	Long-term Vulnerability Reduction	Fatality Management Services	Public Health and Medical Services	Infrastructure Systems	Economic Recovery	
IED Attack: a lone actor deploys "an" improvised explosive device (IED) in an indoor concourse of a stadium during a sporting event	67,500 spectators 2,500 vendors and employees	2,500 vendors and employees	Reinforce 500 concrete support columns in stadium concourse	52 fatalities	350 casualties	N/A	$14 million of direct economic loss (ticket sales, hotel stays, parking, food, and souvenirs)
Accidental Chemical Material Release: A nighttime accident in the rail yard results in the release of a toxic inhalation hazard (TIH) in a densely populated residential area	N/A	350 rail yard employees and first responders	Reroute 100% of rail carrying TIH cargo around densely populated areas	4 fatalities	75 casualties	Damage and contamination to 3 lines at the rail yard	$11 million of direct economic loss (loss of the chemical, phsyical damage to the train, damage to the rail yard)
Earthquake: A magnitude 7.2 earthquake centered near an urban area occurs during mid-afternoon in March	N/A	N/A	Undertake seismic retrofit measures at all public stadiums	375 fatalities	8,400 casualties	350,000 customers without power	$8.4 billion of direct economic loss

Figure 3.4 Sample THIRA analysis.

The THIRA methodology and Core Capabilities are designed as a community-based approach to "provide context and establish capability targets" that are relevant for the given jurisdiction, in other words right-sized, but still standardized according to this methodology. Some US states use this methodology and require its use as part of the Emergency Management Accreditation Program (EMAP)[15] and for emergency management program grants.

BUSINESS CONTINUITY – RISK ASSESSMENT AND BUSINESS IMPACT ANALYSIS

Emergency and public safety managers who are reading this book may wonder why business continuity is included. While emergency management focuses on the three priorities of life-safety (reduce the loss of life and long-term injuries), asset protection (minimize property loss and damage), and stabilizing the incident, business continuity differs as it has a much narrower focus on an enterprise's (used to cover all types of organizations) people, processes, and technology and property following a major disruption (may be a disaster, prolonged power outage, or cyber-attack), with the aim of keeping essential business functions operating while recovering fully from that disruption. Business continuity practices are enterprise-based, which may be a for-profit, non-profit, or other large organization. This contrasts with the broader community-based approach of the THIRA methodology just covered, although business continuity for government is often referred to as COOP for continuity of operations or in the case of catastrophic events continuity of government (COG). However, governmental emergency management agencies, the private sector, non-profit organizations, community, and faith-based organizations all need to work together effectively in what is termed a "Whole Community"[16] approach, familiar to most emergency managers who have been in the field.

Enterprises engaged in analyzing vulnerabilities to risk events may start by looking at the following areas to address internally for any physical assets that may be impacted:

- All computer devices: desktops, laptops, mobile devices, etc.
- Data and telecommunications
- Information systems: local and/or cloud-based servers, networks, and applications; and backup servers and storage
- Hard copy records, both on and off-site
- Logistics chain: warehouse, storage, shipping, transportation
- Manufacturing/production facilities
- Workspaces (company offices and remote work)

In addition to physical assets and systems, business continuity practice also looks at a wider set of factors after a major disruption: employee competencies to respond and recover (do they know where to go and what to do?), upstream impacts (supply chain/vendors), downstream impacts (distributors, brokers, clients/customers), and other external factors (subcontractors/contractors, partners) and environmental limitations (the critical infrastructure areas identified above).

Following the process of risk analysis, which is similar in nature to that in the THIRA methodology, the next step in standard business continuity is called the Business Impact Analysis (BIA). The purpose of a BIA is to identify the critical functions, their vulnerability to impacts, downtime (MTD[17]), and potential ramifications to the enterprise (losses, impacts to clients, etc.). For some industries the maximum tolerable downtime for a specific business function may be very short, for example seconds (financial sector) to minutes or hours (healthcare). A sample BIA chart is given in Table 3.4 as a reference to what this might look like after the impact to a large warehouse operation for a major fruit and vegetable wholesale company (note – this is not inclusive of all measurement of factors such as financial, legal, employees, and reputation). The scenario assumes limited damage to the facility, but some damage outside and to systems (power outages, roads closed).

DISASTER RISK REDUCTION

The World Conference on Disaster Risk Reduction (DRR), held in Sendai Japan in 2015 and endorsed by the UN General Assembly, produced the Sendai Framework for DRR. [18] This framework grew out of the impacts from catastrophic disasters such as the 2004 Indian Ocean earthquake and tsunami and the Sendai Earthquake and Fukushima Disaster in 2011. The Sendai Framework is aligned with the Paris Agreement on climate change and ultimately the Sustainable Development Goals. It advocates for the national governments who are committed to engage in:

Table 3.4 Business Impact Analysis Report after a Major Hurricane

Business Function	Key Processes	IT Dependencies	Criticality (Mission Critical-Vital-Important-Minor)	Max. Tolerable Downtime	Recovery Point Objective (Backlog of Work-only IT)	Key Staff Roles	Resources Impacted	Service Level Agreements/ Contractors
Warehouse operation	Receiving	Inventory management system, vendor management system	Mission critical	6 h	24 h	Shipping/ receiving Mgr.	Loading docks, truck staging/ parking	NA
	Inventory management and reporting	Inventory management system	Mission critical	6 h		Inventory Mgr.	None anticipated	NA
	Order picking	Inventory management system	Mission critical	6 h		Inventory Mgr.	None anticipated	Contract staffing
	Shipment dispatch	Inventory management system, shipping management system	Mission critical	6 h		Shipping/ receiving Mgr.	None anticipated	NA
	Safety/health monitoring	Risk/ESH system	Important	3 days	5 days	ESH manager – Corp.	None anticipated	NA
	Maintenance and cleaning (facility and equipment)	None	Minor	3 days	NA		Equipment supplies	Maintenance vendor

"The substantial reduction of disaster risk and losses in lives, livelihoods and health and in the economic, physical, social, cultural and environmental assets of persons, businesses, communities and countries".

The Sendai Framework on DRR given in Table 3.5 outlines seven global targets to be achieved by 2030.

The DRR methodology describes "hazardous events and disasters as the outcome of continuously present conditions of risk; risk= hazard x exposure x vulnerability". It makes the claim that "there is no such thing as a natural disaster, but that disasters follow natural hazards"[19] and that they are a function of where humans choose to settle and the risks they face due to the combination of hazards, exposure, and vulnerability; and that any losses, both in human lives, health, and financial impacts are a result. This approach is clearly a global one and is intended to be inclusive of developing nations where in the past decades we have seen a greater impact and lack of capability to respond and recover. In keeping with the Sustainable Development goals set out by the UN, it is intended to create a proactive "risk-informed systems-based approach to" investment and development (Figure 3.5).

Table 3.5 Global Targets from the Sendai Framework for Disaster Risk Reduction

Substantial Reductions	Substantial Increases
1. Reduce global disaster mortality	1. Increase the number of countries with national and local disaster risk reduction strategies
2. Reduce the number of affected people globally	
3. Reduce direct economic loss in relation to GDP	2. Substantially enhance international cooperation to developing countries
4. Reduce disaster damage to critical infrastructure and disruption of basic services	3. Increase the availability of and access to multi-hazard early warning systems

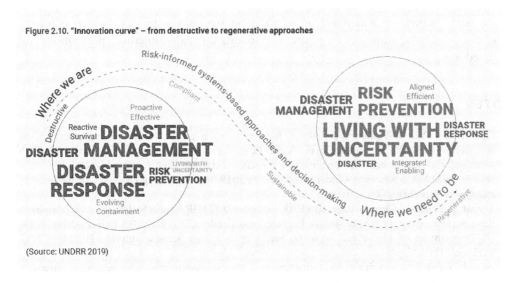

(Source: UNDRR 2019)

Figure 3.5 Innovation curve from destructive to regenerative approaches. (From UN Disaster Risk Reduction 2019.)

WHY DO EMERGENCY MANAGERS AND PUBLIC SAFETY PROFESSIONALS NEED TO BE AWARE OF DISASTER RISK REDUCTION?

As I laid out in the preface of this book, with the rising pressure of climate change, it makes greater sense to focus attention on investment in mitigation and prevention in risk reduction vs. post-disaster risk response after the fact with emergency deployments and recovery spending. A second reason is that, increasingly, professional project managers and departments in large jurisdictions are being called upon to engage in risk reduction projects. For example, the New York City Emergency Management Department (NYCEM) has a unit devoted to hazard mitigation that works with the Department of City Planning (DCP) and the NYC Mayor's Office of Resiliency, and a variety of other NYC agencies.[20] The California Office of Emergency Services (CalOES) has two units that focus on mitigation planning, on both State and local mitigation planning.[21] CalOES leads the effort for the State Hazard Mitigation Plan (SHMP). Lastly, with a growth mindset approach, we can learn a great deal from examples around the world in DRR and improve our comprehension by understanding this context. We will expand on this in the next chapter and revisit this approach later in this book.

CONCLUSION

Any jurisdiction or organization needs to base its emergency management and public safety planning and capacity building with a clear lens on risks and understanding their ramifications. In this chapter we discussed the foundation for assessing, prioritizing, and analyzing the risks that arise from threat and hazard events. We explored these first major steps of the risk management process and covered different approaches from governmental to commercial sectors; this included defining the dynamics of these risk events, identifying the categories of threats and hazards and different potential areas of exposure. In the next chapter we will move on to the next step in the process, which is identifying and developing strategies for addressing threats through mitigation, prevention, and preparedness.

NOTES

1 **Operational Period:** The time scheduled for executing a given set of operation actions, as specified in the IAP. Operational periods can be of various lengths but are typically 12–24 h (may be shorter or longer) for faster moving events such as wildfires, tornados, or hurricanes. Source: FEMA ICS Resource Center as of June 6, 2019 https://training.fema.gov/emiweb/is/icsresource/glossary.htm#O

2 Threat and Hazard Identification and Risk Assessment (THIRA) and Stakeholder Preparedness Review (SPR) Guide; Comprehensive Preparedness Guide (CPG) 201, 3rd Edition, May 2018. US Department of Homeland Security. See: www.fema.gov/sites/default/files/2020-04/CPG201Final20180525.pdf

3 ISO 22310:2019 Security and resilience – Business continuity management systems – Requirements. See www.iso.org/standard/75106.html

4 NFPA 1600 Standard on Continuity, Emergency, and Crisis Management. See www.nfpa.org/codes-and-standards/all-codes-and-standards/list-of-codes-and-standards/detail?code=1600

5 Hillson, D. (2000). Project risks: Identifying causes, risks, and effects. PM Network, 14(9), 48–51. See www.pmi.org/learning/library/project-risks-causes-risks-effects-4663

6 History – Ferdinand de Lesseps, source: www.bbc.co.uk/history/historic_figures/de_lesseps_ferdinand.shtml

7 A number of definitions for risk management can be found in the glossary.

8 National Weather Service 10th Commemoration of Joplin, Missouri EF-5 Tornado. See https://storymaps.arcgis.com/stories/d64c8d41c28949bd87478f252ac7e378

9 Online FEMA Glossary, derived from Acronyms Abbreviations and Terms a Capability Assurance Job Aid, FEMA, P-524/July 2009 See https://training.fema.gov/programs/emischool/el361toolkit/glossary.htm#H. Also, references FEMA 2011 The Strategic National Risk Assessment in Support of PPD 8: A Comprehensive Risk-Based Approach toward a Secure and Resilient Nation. See www.dhs.gov/xlibrary/assets/rma-strategic-national-risk-assessment-ppd8.pdf

10 E. L. Quarantelli (ed.): What Is a Disaster?. Natural Hazards 18, 87–88 (1998). https://doi.org/10.1023/A:1008061717921

11 National Hurricane Center NOAA www.nhc.noaa.gov/aboutsshws.php

12 President Policy Directive-21 (PPD-21) issued in 2013 "advances a national unity of effort to strengthen and maintain secure, functioning, and resilient critical infrastructure". See www.cisa.gov/critical-infrastructure-sectors
   FEMA created Community Lifelines in 2018 as a way "to reframe incident information, understand and communicate incident impacts using plain language, and promote unity of effort across the whole community ...during incident response" and were incorporated into the National Response Framework in 2019.

13 Mission Areas and Core Capabilities – The National Preparedness Goal identifies five mission areas and 32 core capabilities intended to assist everyone who has a role in achieving all of the elements in the Goal. www.fema.gov/emergency-managers/national-preparedness/mission-core-capabilities

14 Threat and Hazard Identification and Risk Assessment (THIRA) and Stakeholder Preparedness Review (SPR) Guide; Comprehensive Preparedness Guide (CPG) 201, 3rd Edition, May 2018. US Department of Homeland Security.

15 EMAP, as an independent non-profit organization, fosters excellence and accountability in emergency management programs, by establishing credible standards applied in a peer review accreditation process. The ANSI/EMAP 5-2019 Emergency Management Standard by EMAP is the set of 66 standards by which programs that apply for EMAP accreditation are evaluated. www.emap.org/

16 A Whole Community Approach to Emergency Management: Principles, Themes, and Pathways for Action. FDOC 104-008-1/December 2011, US Department of Homeland Security, FEMA. See www.fema.gov/sites/default/files/2020-07/whole_community_dec2011__2.pdf

17 MTD= maximum tolerable downtime. This is the total time that a business function can be down before it has a significant financial or critical impact to the enterprise.

18 What is the Sendai Framework for Disaster Risk Reduction, United National Office for Disaster Risk Reduction? www.undrr.org/implementing-sendai-framework/what-sendai-framework

19 Understanding Disaster Risk www.preventionweb.net/understanding-disaster-risk/component-risk/disaster-risk

20 New York City Hazard Mitigation https://nychazardmitigation.com/about/

21 California OES Hazard Mitigation Planning www.caloes.ca.gov/cal-oes-divisions/hazard-mitigation/hazard-mitigation-planning

Chapter 4

Developing Strategies and Capabilities to Manage Major Risk Events

Climate change adaptation projects will proceed as planned in the business-as-usual scenario, but retreat and relocation will increasingly be coping strategies of choice, as life in coastal areas and other disaster-prone zones become more and more untenable. Property values will plummet in hazardous areas, and insuring at-risk properties will become prohibitively expensive or impossible at some point, wiping out homeowners' investments. Jurisdictions worldwide will face the challenge of extensively retrofitting, relocating, or abandoning critical infrastructure elements such as ports, highways, and power plants. Distinctive local and regional cultures threatened by disasters and climate change, such as cultures that are dependent on fishing and other subsistence activities, will weaken or disappear.

*Kathleen Tierney, Director Emerita, Natural Hazard Center; Professor Emerita, Department of Sociology, University of Colorado Boulder (2019). Author of **Disasters: A Sociological Approach**. United Kingdom: Wiley. p. 227*

"An ounce of prevention is worth a pound of cure" so the saying goes, and this is proving increasingly true in this age of global warming and the catastrophic impacts of natural hazards being wrought on communities. We find few communities untouched by natural hazards with frequent, intense historic wildfires in Australia, the American West, and in once rare areas such as Russian Tundra to epic storms, hurricanes, and typhoons globally. We hear the phrases "unprecedented" or "unexpected" in the aftermath of extreme hazardous events so often now that it has become perfunctory. In this chapter, we will discuss the types of hazards and address the options for risk management strategies. As laid out in the previous chapter, it is important to keep in mind that each jurisdiction and organization will have different considerations based on its unique geography, capabilities, and access to resources.

Building from our initial risk assessment and analysis phase in which we have scanned the environment, identified risks (threats and hazards), and analyzed them by providing context, defining the potential impacts subject to exposures and vulnerabilities, we will now move to the next step in the process: treating risks. Treating risks encompasses looking at the vulnerabilities, reviewing existing protective and mitigation measures, reviewing any residual risk exposures, and then identifying viable and practical options to address their causes and impacts, see Figure 4.1.

The options identified as prevention, mitigation, or preparedness measures, which will be discussed below, may produce project opportunities which then begins with the project

DOI: 10.1201/9781003201557-5

Figure 4.1 Risk management process. (From Elliott, Michael W. *Risk Management Principles and Practices*, 2nd Edition (The Institutes 2016), p. 5.16; used with permission from the American Institute for Chartered Property Casualty Underwriters.)

development process covered in the following chapters. Some examples of these types of projects are presented below.

IDENTIFYING STRATEGIES TO ADDRESS RISK EVENTS: MITIGATION, PREVENTION, AND PREPAREDNESS

Let's create some context for this section by starting with the Bowtie Analysis from the previous chapter as a reference example, a river flooding event in a town that would inundate the flood plain. Some basic options for mitigation, preparedness, and response are provided in Table 4.1.

Let's explain what each of these major actions means and how it is applied. As stated previously in this book, this chapter is intended to provide enough grounding in this topic for readers who are new to the fields of emergency management and public safety and not intended to go into great depth on the topic of risk. Also, note that we are focusing on a natural hazard event in the example provided, so the area of prevention is not covered at any length using this example, as it is in the domain of criminal or terrorist acts covered under public safety. While we address how project management may be practiced in a public safety setting, and, certainly the practices, examples, and cases presented here are applicable to public safety projects, it is not the focus of this book. Lastly, FEMA's Core Capabilities,[1] part of the US National Preparedness Goal, include five mission areas: prevention, protection, mitigation, response, and recovery.[2] The missions of protection, (emergency) response, and recovery are mostly focused on post-event response and are therefore addressed largely in Chapter 12 Emergency Response as a Project. This chapter deals with pre-event disaster mitigation and preparedness.

Table 4.1 Bowtie Analysis of a Flood Event

Causal Factor(s)	Prevention/Mitigation/ Preparedness Measures	Event	Primary Mitigation/First Response Measures	Direct Consequence(s)	Recovery Actions
Heavy rain creates riverine flooding which inundates the town flood plain	**Community Based:** • Zoning restrictions (limits on building) • Flood mitigation design and engineering • Building berms/ purchasing flood walls/barriers • Evacuation plans • Plans for emergency shelters **Organizational (offices/ facilities in the flood zone):** • Flood walls • Backup worksite • For IT-Cloud-based servers and systems, spare laptops • Reciprocal sites (production)	Houses/ business in the flood plain flooded with 3 ft. of water	**Community Based:** • Sandbags/flood barriers • Evacuation plans (orders and transportation) • Establishing emergency shelters **Organizational:** • Alerting employees – work from home, backup, and reciprocal sites • Activating backup and reciprocal sites	**Community Based:** • Foundation and structural damage to homes • Electrical outage • Sewage backup • Property damage (furniture/ appliances) mold growth • Drowning • Infection from polluted water **Organizational:** • Employees impacted/ unavailable • Transportation to facilities disrupted • Building in flood zone flooded	**Community Based:** • Debris removal • Restore electricity/ water • Damage assessment/repair • Mud-out homes • Clearing storm drains • Coordinate and facilitate public assistance and aid programs **Organizational:** • Repair damaged facilities • Restore primary servers and IT systems • Support staff recovery

Note: The example is intended as notional and not meant as an in-depth analysis. Some additional examples are provided in this chapter.

PREVENTION

As you might expect, prevention is defined as the action taken to avoid a disaster incident or intervene to stop an incident from occurring. This is a US Department of Homeland Security (DHS) terrorism-focused component that became part of the phases of emergency management following the events of 9/11. It was developed to differentiate funding streams separately from mitigation, which is intended to address impacts from mostly natural hazards and technical accidents or cyber-attacks. Much of this centers around the actions taken for what are commonly titled "CBRNE" events: attacks that would involve chemical, biological, radiological, nuclear, and explosive device elements. Prevention-related funding was made available to jurisdictions to harden their infrastructure from acts of to terrorism, and police departments sought funding to largely procure additional equipment and new surveillance-based technologies, and to a lesser degree to bolster their staff, training, and other resources.

FEMA defines seven core capabilities for prevention. These are listed in Table 4.2 along with some examples of potential projects associated with each:

Table 4.2 Core Capabilities for Prevention and Project Examples

Core Capability	Project Examples
Planning	Planning a bio-hazard exercise of a new plan
Public information	Adopting a mass notification system for emergency alerts in an organization for any terrorism-related incidents[1]
Operational coordination	Adopting a secure system for intelligence gathering for a common operating picture under a Fusion Center[2]
Intelligence and information sharing	Implementing a license plate screening system at a law enforcement agency (state/local police, investigative agency)
Interdiction and disruption	Implementing radiological detection equipment (rad detectors) and systems
Screening, search, and detection	Installation of a monitoring system and security cameras externally and internally around a facility
Forensics and attribution	Implementing a tracking system for collected bio-medical samples

www.dhs.gov/fusion-centers

[1] Of course an emergency alert system has purposes beyond just terrorism incidents.
[2] Fusion Centers are state-owned and operated centers that serve as focal points in states and major urban areas for the receipt, analysis, gathering and sharing of threat-related information between state, local, tribal and territorial (SLTT), Federal and private sector partners.

Building Counterterrorism Capabilities at FDNY After 9/11[3]

The events of 9/11 shook the New York City Fire Department (FDNY) to its core. The FDNY lost 343 firefighters and leadership staff that day, not to forget those whose lives were cut short by the effects of lung disease and mental illness in the years after. Its command/control and communications infrastructure were severely challenged and pointed to major gaps. Following this catastrophic loss and lessons learned from thorough after-action reviews internally and externally, the agency leadership took on the task of building out capabilities in incident command and counter terrorism, which did not exist before then.

In 2004 the FDNY established the Center for Terrorism and Disaster Preparedness (CTDP), devoted to developing state-of-the-art practices and applying them in the organization through planning, training, and exercises. The CTDP was designed to look at the wide range of major threats, from natural hazards to potential acts of terrorism, and how the agency would respond and coordinate its work. This included looking at how FDNY integrates with other City, Federal, and State agencies, as well across jurisdictions. ICS was one of the cornerstones of this new capability, which made this a core element of the agency's command structure.

Since the initiation of the CTDP and the development of these new ICS capabilities, FDNY was able to provide assistance to the City of New Orleans following Hurricane Katrina and help coordinate efforts among a number of different support agencies. In the wake of Hurricane Sandy in 2012, FDNY was called upon to stand up a command post and staging area at a former air-base in Brooklyn, Floyd Bennett Field, to support the recovery efforts and coordinate them across dozens of first response agencies for residents impacted.

Table 4.3 Core Capabilities for Mitigation and Project Examples

Core Capability	Project Examples
Planning	Land use planning to address flood prevention, fire hazard, seismic impacts, and environmental and community impacts. Building codes and financial incentives (buyouts and bonds)
Public information and warning	National Weather Service or National Hurricane Center weather prediction systems upgrades. Implementation of IPAWS at the local level
Operational coordination	Developing coordinated community evacuation plans (for flood, fire, hurricane hazards)
Community resilience	Purchasing hazard insurance (flood, wind, fire).[1] Contracting for catastrophic bonds.[2]
Long-term vulnerability reduction	Implementation of structural flood controls (berms, barriers, breaks)
Risk and disaster resilience Assessment	Hazard mapping (as mentioned before: HAZUS, NFIP, USGS, GIS)
Threat and hazard identification	Conducting a threat and hazard analysis (see Chapter 2; in principle this activity belongs in the previous chapter, but FEMA considers this to be a part of the overall mitigation process)

[1] Insurance is included here although I look at it as only a cost mitigation measure, as it does not reduce the physical impact or probability of an event happening, and the payments for claims come after an event.
[2] "A CAT bond is a security that pays the issuer when a predefined disaster risk is realized, such as a hurricane causing $500 million in insured losses or an earthquake reaching a magnitude of 7.0". www.chicagofed.org/publications/chicago-fed-letter/2018/405

MITIGATION

Mitigation is the "sustained action to reduce or eliminate long term risk to life, property, and environment from natural and human caused hazards".[4] Mitigation actions are intended as long-term and enduring solutions that reduce the impacts of hazardous threats significantly or fully, thereby reducing the reliance on response or recovery efforts. The driving reasons for investment in mitigation measures are that the costs for response and recovery outweigh the costs for mitigation. Following major disaster events that have occurred in the United States, such as the 1906 San Francisco Earthquake,[5] the 1991 Oakland Hills Conflagration,[6] and Hurricane Andrew in 1992, efforts increased to implement mitigation measures in those communities, and at different government levels, local, state/tribal, and Federal.

In the past, emergency managers have been largely focused on emergency response and recovery activities. Considering this expansion for emergency management as a whole, those in the emergency management profession have been called on to play a large role in planning and, in many cases, leading mitigation programs.

FEMA defines eight core capabilities for mitigation. These are listed in Table 4.3 along with some examples of potential projects associated with each.

US FEDERAL FUNDING PROGRAMS FOR HAZARD MITIGATION

At the Federal level, going back to 1997 with the introduction of Project Impact, FEMA has paid greater attention to investing in mitigation-based projects. Greater emphasis was put on state, tribal, and local jurisdictions to develop mitigation plans as a result of the Disaster Mitigation Act (DMA) of 2000.[7] While adoption was slow at first, this has increased over the past two decades. Listed below are some of the most significant Federally funded mitigation programs in the United States:

Hazard Mitigation Grant Program	Community Assistance Program
Pre-Disaster Mitigation Program	National Earthquake Hazard Reduction Program
Flood Mitigation Assistance Program	National Hurricane Program
Repetitive Flood Claims Program	National Dam Safety Program
Severe Repetitive Loss Program	Fire Prevention and Assistance Act

From the program titles, you can see that these grant-funded programs cover a wide range of natural and technological hazard types. Some of these programs require matching funding at the state and local level.

Resilience by Design: Earthquake Mitigation in Los Angeles[8]

Los Angeles (LA) is no stranger to earthquakes, with the 1994 Northridge Earthquake recorded as the last major quake, striking LA's San Fernando Valley area[9] which resulted in more than 60 dead, 9,000 injured, and caused damage amounting to over $20 billion. Learning lessons from this and other seismic events and earthquake impact modeling, the City of LA developed a set of recommendations in 2014, culminating in a report *Resilience by Design*. The report used a simulated 7.8 magnitude earthquake which modeled impacts that would result in potentially thousands dead and over $213 billion in economic losses.

The recommendations addressed vulnerabilities from earthquakes with practical, cost-effective solutions targeting several key areas of seismic vulnerability:

- pre-1980 "non-ductile reinforced concrete" and "soft-first-story" (primarily wooden) buildings;
- water system infrastructure (including impact on firefighting capability);
- telecommunications infrastructure.

A primary vulnerability to seismic impacts were residential buildings, accounting for over 16,000 buildings in LA. [10] This report resulted in a new City ordinance which required a complete retrofit or demolition of

- wood frame buildings with ground floor open, front wall lines such as a parking garage or store front, 2+ stories built prior to 1/1/1978, with exemptions for three or fewer residential units;
- concrete buildings with 1+ stories built prior to 1/13/1977, with exemptions for detached single and duplex family homes.

The law required that owners submit proof of a retrofit or plans for retrofit or demolition, obtain a permit for the same within 3.5 years, and that the work would be completed within 7 years. While it was controversial when it was passed, in light of funding primarily coming from building owners and renters, as of 2021 93% of buildings met the requirements or were on their way to completing their plans for the seven-year deadline.[11]

PREPAREDNESS

In emergency management, *Preparedness* is defined as "the action to strengthen a community's ability to respond effectively to an event". Key emergency preparedness activities include planning, training, and exercises (includes drills, table-top exercises or TTX, functional, and full-scale exercises). While emergency preparedness is part of the cycle of emergency management as we discussed in Chapter 2, FEMA does not include this as a distinct mission in its core capability model.

Emergency management professionals whose work is dedicated to emergency response (not as part of first response with fire, law enforcement, or EMS) will often acknowledge that 90% or more of their time is taken up with preparedness activities, although this is changing as we discussed earlier. The typical cycle is as given in the order presented in Figure 4.2, in which emergency management staff: develop, review, and revise the plan; train staff on the plan; develop and conduct exercises that evaluate emergency response based on the plan and training and seek to improve it. This is basically the continuous improvement cycle: plan, do, check, act, or the PDCA.

There are a multitude of emergency response plans, some all-hazard or others that are hazard specific, depending on the type of organization and the types of hazards and risks they face. FEMA offers different options for emergency management planning as part of the National Preparedness Goal (see: www.fema.gov/emergency-managers/natio nal-preparedness/plan). The most common guide for governmental jurisdictions is the Comprehensive Preparedness Guide 101 (CPG v.3, issued in 2021), see: www.fema.gov/ sites/default/files/documents/fema_cpg-101-v3-developing-maintaining-eops.pdf.

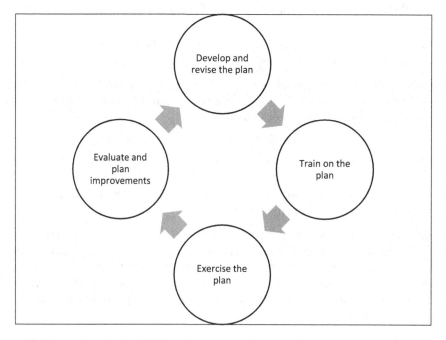

Figure 4.2 Plan–train–exercise (PTE) cycle.

The most common approach is to develop what is more generally referred to as an Emergency Operations Plan or EOP, or more specifically a Comprehensive Emergency Plan or CEMP, which is an all-hazards approach, and then develop incident or hazard-specific annexes (hurricane, wildfire, terrorism, cyber security, power outage, etc.).

When it comes to guides on training, FEMA again offers a range of resources and centers that support different levels and needs for emergency preparedness training, see: www.fema.gov/emergency-managers/national-preparedness/training. This includes access to online training, institutions of higher education, and grants to support plan and jurisdiction-specific training needs. You will find some limited guidance and templates concerning training development later in this book.

The Homeland Security Exercise and Evaluation Program (HSEEP) addresses the wide range of exercise types and includes workshops which may be used for plan development. See: www.fema.gov/emergency-managers/national-preparedness/exercises/hseep

As we discussed earlier, each element given above, developing an emergency plan, training, or an exercise, can each be considered a project, regardless of its size. However, the complexity of the management methods and tools should be chosen and adapted to suit the project, as we will explore in further chapters.

RESILIENCE VS. MITIGATION AND PREPAREDNESS

The term "resilience" has come into common usage around emergency management, disaster response, and recovery but is often hotly discussed among those in risk, emergency, and public safety management as to its actual meaning. For our purposes, we will define resilience as:

> The ability for a community (town, city, or region) to return to a reasonable level of its previous normal economic level and type of activities (e.g. back to business, daily routines) within an acceptable period of time following a major disaster event or disruption.

In this context, mitigation and preparedness mitigation are focused on reducing or eliminating risk impacts and would be considered a subset of resilience activities. Preparedness, as we presented above, is focused on responding to risk impacts from a disaster.

IDENTIFYING OPTIONS FOR RISK STRATEGIES: AVOIDANCE, ACCEPTANCE, MITIGATION, AND TRANSFERENCE

Avoidance

Now, avoiding or accepting risks is usually not feasible, especially for major risks. We can only truly avoid a risk event by not exposing ourselves to the risk. Individuals, households, and, at times, whole communities may decide to pick up and move away from potential hazard risks. Hurricane Katrina affected the demographic make-up of New Orleans, Louisiana, and we witnessed a major diaspora of households which relocated to Houston, Texas, as well as other parts of the United States.[12] In fact, about 95 million US residents live in coastal areas or roughly 29% of the US population, with about 60 million who live in areas vulnerable to hurricanes.[13] A third of all US housing units are located in the wildland urban interface (WUI), with California, Texas, and Florida, as the three top states with the greatest vulnerability by population to wildfire risk.[14] Two factors have increased the probability and severity of wildfire risks: building in areas within the WUI; and global warming, as evidenced by record-breaking heatwaves, and some of the most costly wildfire events. The bottom line here is that the trend is not toward relocating away from risky areas, but in accepting more risk.

Acceptance

On the other end of the spectrum of risk response, we have acceptance. The acceptance strategy, as it sounds, accepts the risk and responds at the time it happens; it does not mean planning for or preparing for the risk event, as that would fall into the category of mitigating the impacts of the risk with preparedness activities or allocating funding for recovery. For the most part, willingly accepting major risk events (hazards and threats) is also not a viable option. This may be a plausible option for dealing with lower stakes risk events or low probability events which may be less frequent or predictable, where rebuilding may be the last resort, such as with major tornados or earthquakes. At least in the United States, we have weighed our options and have tended to lean away from merely accepting risks to more actively managing risks through mitigation and transference options.

Mitigation

The mitigation option was discussed at greater length above, but it is important to note that this option and the transference option are not mutually exclusive, and typically used in conjunction, looking at the mix of risk reduction, or what is commonly referred to as paying down risks through active mitigation measures and then transferring the residual risks through insurance. We can look at this, similar to how this is viewed by the risk management industry as the "Lines of Defense" model, as illustrated using a wildfire event in Figure 4.3.

Transference

Risk transference through the purchase of insurance or a similar financial instrument or contract, such as CAT bonds, is a reactive risk management measure. It is reactive in that the insurance policy holder, whether a household or organization, takes a loss and then makes an insurance claim against the policy in response *after* a risk event has taken place, not before or during. The claim is then reviewed, along with a damage transferring the risk impacts to a third party (insurance company) should only be considered the last line of defense after mitigation measures have been analyzed and deemed insufficient to cover any substantial residual risk exposure or too expensive to pursue further.

Hazard-based insurance covers all types of losses from damage to buildings, infrastructure, hard assets (equipment), supplies and materials, as well as business interruption and

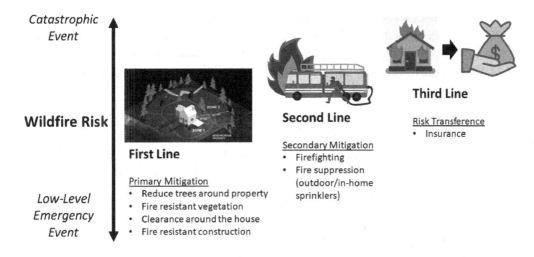

Figure 4.3 Wildfire risk and three lines of defense. (Image sources: CalOES defensible space flyer; FEMA firefighter wildfire herographic; Rawpixel CC01.0 Universal (CC01.0) Public Domain Dedication).

coverage for homeowners. Both for commercial enterprises, governments, and building owners/homeowners, the most common lines of commercial insurance cover hazard risks from:

- Flooding (may be different types: rain, flood, storm surge)
- Wind (hurricane and tornado)
- Fire/smoke (accident, wildfire)
- Earth movement, not limited to earthquake
- Hail/sleet
- Cybersecurity
- Terrorism
- Active shooter incidents (which includes liability, not a part of this discussion)

Insurance plays a major role in recovery from a disaster event when mitigation measures have failed. Besides the commercial insurance marketplace, the US Federal Government and some US States offer subsidized or supplementary funding for insurance to cover flood, fire, or wind damage risks in areas where commercial insurers may be unwilling to provide coverage. The National Flood Insurance Program (NFIP) is the most well-known of these programs. The NFIP was established in 1968 under the management of FEMA to help fund insurance for structures in what is termed by FEMA as the Special Flood Hazard Area or the 100-year flood plain, where the risk of flooding is considered to be 1% in any given year.[15]

Well intended from its inception, the increase in levels and frequency of storm-related flooding, whether caused by heavy rain, rivers, or storm surge, has caused the program to go into steep debt to the tune of over $20 billion to the US Government. This was in large part due to Hurricanes Katrina, Rita, Ike, and Sandy between 2005 and 2012. Between 2010 and 2019 there were 123 disaster events with an estimated $858 billion of damage, with over $300 billion in 2017 alone.[16] The majority of this damage is storm related.

The flood maps developed by FEMA under the NFIP are updated periodically and have expanded the designated flood zones, resulting in push-back from local communities, many of them in lower socio-economic categories who cannot afford to pay for the additional insurance premiums.[17] There is also an ongoing debate around how the NFIP program is managed, funded, and whether it should better reflect real market rates.[18] However, many communities are underinsured against hazard risks such as fire or flooding. Buildings, including some homes, are not required to carry flood insurance. Less than 25% of the buildings inundated by Hurricanes Harvey, Sandy, and Irma had flood insurance coverage.[19]

In addition to the NFIP, in the wake of any Federally declared disaster under the Stafford Act, FEMA offers assistance to individuals, households, businesses, organizations, and governments. Disaster assistance is offered based on a formula for the estimated costs of disaster impacts largely based on population, is means based, has a cost sharing with state governments (usually 25%, but this is sometimes waved based on the severity of the disaster), and is not intended as a substitute for commercial insurance.

While the FEMA Assistance Program is a well-known feature in Federally declared disasters, payments are often slow to make it through to recipients and are channeled through state agencies, which can create another layer of bureaucracy, so delays are common in compensating disaster survivors who make claims through this program. While FEMA has historically made it clear that this program is designed to build back to a previous state before the disaster, in the past two decades this restriction has changed to help jurisdictions build back facilities, homes, or infrastructure to a more resilient state or to help relocate individuals or whole communities through "buy-outs".

As with previous chapters and subsections, there are many resources on the topic of risk transfer and FEMA programs, and this section of the book is not intended to exhaustively cover the topic of insurance. It is highly recommended that you consult with a risk management expert when exploring any coverage options as part of a risk management strategy.

COST-BENEFIT ANALYSES

The final stage in planning a risk response strategy is reviewing the viable options and conducting a cost-benefit analysis to identify an optimal mix or selection of projects as part of a risk response strategy. Choosing which projects to invest in can be part of a portfolio management process, which we will address in Chapter 14.

Making sure you have identified and collected all of the associated costs and benefits or value derived from different options can be challenging. Some sources of costs and benefits:

- Historical losses for property/facilities/infrastructure or comparable assets (all fixed property and hard assets – furnishings, equipment, wired systems, etc.)
- Total value at risk (total cost of assets)

Table 4.4 Hurricane Scenario and Risk EMV Analysis

Hurricane Scenario	Probability	Estimated Cost of Damage	Expected Monetary Value
Scenario A. Category 1	10%	$20 million	$2,000,000
Scenario B. Category 2	5%	$40 million	$2,000,000
Scenario C. Category 3	1%	$100 million (near total loss)	$1,000,000
		Total EMV for Scenarios A–C	$5.000,000

- Expected monetary value (EMV) based on a range of scenarios[20]
- Obtaining estimated costs through experts for specific options, such as engineering mitigation measures or insurance coverages

Once you have pulled together all related costs and potential losses, then you can develop a calculation for the EMV based on the probability. An example of this is provided below for a coastal community looking at putting in sand dunes, complete with sea grasses,[21] to help protect homes and businesses in a flood zone, as well as reduce beach erosion:

$$\frac{\text{Benefit}}{\text{Cost}} = \frac{\text{Loss from flood damages in value of fixed assets-homes and businesses}}{\text{Project cost to create dunes}}$$

$$\frac{\text{Benefit}}{\text{Cost}} = \frac{\$100 \text{ million}}{\$15 \text{ million}} = \text{Note – we are missing potential loss and probability.}$$

But we are not done yet, as we need to look at the probability and potential damage from different flooding events. Since there is a range, we will use the EMV formula = probability × value of the outcome, in this case the potential cost in potential losses. We provide three different possibilities based on potential hurricane strengths categories 1–3 in Table 4.4.

It is important to take into account that $5 million is the EMV for one year for this limited set of potential hurricane events. Over the course of a ten-year period, the probability of at least one category 1 storm hitting this coastal community is 10 × 10% or 100%, so if we return to our formula and calculate this based on a ten-year span, we come to the following conclusion:

$$\frac{\text{Benefit}}{\text{Cost}} = \frac{\$50 \text{ million}}{\$15 \text{ million}} = \text{Over a ten-year period; 10 years} \times \$5 \text{ mill. EMV} = 3.3 \text{ times benefit to cost}$$

We can do the same for insurance coverages or other types of mitigation measures. Much of this cost-to-benefit ratio depends on a range of factors we should also consider part of our basic cost estimates:

- the time horizon for the investment (as well as any ongoing maintenance costs);
- the actuarial science of probability and potential cost of loss from damages, taking into account any other factors such as rise in sea level or changes in frequency or severity of hurricanes;

- changes to the built environment, including any new building;
- other mitigation measures taken by the homes and businesses in the area;
- other intangible value/benefits or disadvantages; these are very hard to analyze and estimate as they may be associated with socio-political concerns of community, for example cultural and historical priorities.

This is a simple example, and for many communities in the United States, as well as elsewhere in the world, funding for mitigation measures is always a challenge. More often than not, this funding comes from Federal (National) programs funneled through state (provincial or regional) emergency management agencies. Even though investments in long-term mitigation projects are difficult to get backing for and sustain, the research and experience indicate that there are savings associated engineered risk mitigation measures, whether for wildfire or flooding. According to a 2019 report from the National Institute for Building Sciences, investment in effective mitigation measures has a savings of $13 for every $1 invested.[22]

We will explore how to develop accurate cost estimates for projects using different estimating methods and how we arrive at them. We have now set the stage in addressing risks, and identifying and selecting potential projects. In our next chapter we will begin from the start in developing a project plan by looking at the scope and stakeholders.

The Story of Indianola, Texas

The town of Indianola, Texas, is long gone but was once a bustling port city on the Texas coast. It was located between Corpus Christi and Galveston and aspired to be what Houston eventually grew to become, a major Southeast Texas port metropolis. It had an ideal harbor, and it would have achieved that status had it not been for two major hurricanes, back to back in the late 1800s. The first one hit in 1875, and the second in 1886, each hurricane delivering a devastating blow to the town. What little was rebuilt from the first hurricane was abandoned by the second.

Now, the town of Galveston further North knew of this history, and the town elders decided to begin building a seawall and raising the grade level starting in 1902, just after they experienced a devastating hurricane in 1900 that killed between 6,000 and 12,000 inhabitants out of a population of 40,000 at the time. Alas, the construction of the seawall was only partially built, and in 1915 another major hurricane hit with a storm surge of 12 ft. and caused considerable damage. The upshot of this event was to boost the resolve of the citizens to finish the seawall, which was finally completed in 1963 at an estimated cost of around $150 million.[23]

By this time, a canal was built to the City of Houston, which outpaced it in population and port growth. Galveston remains a major resort town to this day, but never measured up to the thriving port city it aspired to, as Houston became, and as it's departed counterpart of Indianola once did. During Hurricane Ike, storm surge topped the seawalls resulting in some, but not major flooding damage. However, this did spur review of whether further mitigation measures were needed for future storms. There are plans to build a coastal storm barrier to protect the Galveston Bay and Texas coast in the greater Houston area, nicknamed the Ike Dike, after Hurricane Ike.[24]

NOTES

1 There are five mission areas and 32 activities that support these mission areas: www.fema.gov/emergency-managers/national-preparedness/mission-core-capabilities

2 There are five mission areas and 32 activities that support these mission areas: www.fema.gov/emergency-managers/national-preparedness/mission-core-capabilities

3 This was summarized from two sources: The FDNYS Preparedness and Training Post 9/11, Diana Kelly Levey www.dianakelly.com/portfolio/the-fdnys-preparedness-and-training-post-911/; and The Leadership Podcast; Heidrick and Struggles, by Cheryl Stokes, with Deputy Commissioner Mike Puzziferri. Preparing for the unexpected: Insights from a 9/11 FDNY deputy fire chief www.heidrick.com/en/insights/podcasts/episode_40_preparing_for_the_unexpected_insights_from_a_911_fdny_deputy_fire_chief

4 FEMA Glossary, definition for Mitigation: https://training.fema.gov/programs/emischool/el361toolkit/glossary.htm

5 While it is called the Great San Francisco Earthquake, most damage resulted from the resulting fire that it caused and the lack of fire hydrants and fire-fighting crews at the time. See: https://earthquake.usgs.gov/earthquakes/events/1906calif/18april/

6 The Oakland-Berkeley Hills Fire: An Overview by Captain Donald R. Parker, Oakland Office of Fire Services www.sfmuseum.org/oakfire/overview.html

7 Carlos Samuel, Laura K. Siebeneck, Roles revealed: An examination of the adopted roles of emergency managers in hazard mitigation planning and strategy implementation, International Journal of Disaster Risk Reduction, Volume 39, 2019,101145, ISSN 2212-4209 https://doi.org/10.1016/j.ijdrr.2019.101145

8 Resilience by Design. www.eeri.org/images/archived/wp-content/uploads/Garcetti-Los-Angeles-Earthquake-Plan.pdf

9 www.lacity.org/highlights/remembering-northridge-earthquake

10 Los Angeles Requiring Earthquake Retrofitting for 15,000 of Its Scariest Apartment Buildings, Jeff Wattenhofer, October 9, 2015. https://la.curbed.com/2015/10/9/9912754/los-angeles-apartments-earthquake-retrofit-law

11 Mayor Garcetti and Optimum Seismic, Inc. Highlight Earthquake Retrofits for Making 7,000 Los Angeles Apartment Buildings Safer from Quakes Stronger Buildings Protect Tens of Thousands of Residents, PR Newswire October 22, 2021 www.prnewswire.com/news-releases/mayor-garcetti-and-optimum-seismic-inc-highlight-earthquake-retrofits-for-making-7-000-los-angeles-apartment-buildings-safer-from-quakes-301406459.html

12 Charts show how Hurricane Katrina changed New Orleans, Hurricane Katrina had lasting effects on the physical and social makeup of the Big Easy. August 29, 2015 www.nationalgeographic.com/science/article/150828-data-points-how-hurricane-katrina-changed-new-orleans

13 About 60.2M Live in Areas Most Vulnerable to Hurricanes, Darryl Cohen, July 15, 2019 America Counts: Stories Behind the Numbers US Census Bureau. www.census.gov/library/stories/2019/07/millions-of-americans-live-coastline-regions.html

14 The Wildland Urban Interface is Growing in the United States, Julian Marks, October 5, 2021 Population Geography, GEOGRAPHYREALM. www.geographyrealm.com/the-wildland-urban-interface-is-growing-in-the-united-states/

15 www.fema.gov/glossary/special-flood-hazard-area-sfha

16 NOAA National Centers for Environmental Information (NCEI) U.S. Billion-Dollar Weather and Climate Disasters (2021). www.ncdc.noaa.gov/billions/, DOI: 10.25921/stkw-7w73

17 Flirting with Disaster: Flood Zones Still Uninsured Years after Sandy, Ann Choi, Christine Chung, Clifford Michel, and Greg B. Smith October 28, 2019, The City www.thecity.nyc/special-report/2019/10/28/21210772/flirting-with-disaster-flood-zones-still-uninsured-years-after-sandy

18 Trump FEMA Chief Backs Reducing Federal Role in Disaster Relief, Flood Insurance, Christopher Flavelle, August 23, 2017 Insurance Journal www.insurancejournal.com/news/national/2017/08/23/462113.htm

19 Moving the Needle on Closing the Flood Insurance Gap Carolyn Kousky, Brett Lingle, Howard Kunreuther, and Leonard Shabman. Issue Brief February 2019 Wharton University of Pennsylvania, Risk Management and Decision Process Center. https://riskcenter.wharton.upenn. edu/wp-content/uploads/2019/02/Moving-the-Needle-on-Closing-the-Flood-Insurance-Gap.pdf

20 Expected monetary value (EMV) analysis is a statistical concept that calculates the average outcome when the future includes scenarios that may or may not happen. An EMV analysis is usually mapped out using a decision tree to represent the different options or scenarios. EMV for a project is calculated by multiplying the value of each possible outcome by its probability of occurrence and adding the products together. www.pmlearningsolutions.com/blog/sensitiv ity-analysis-versus-expected-monetary-value-emv-pmp-concept-26#:~:text=Expected%20m onetary%20value%20(EMV)%20analysis%20i

21 This is a fictional example and the numbers are not based on actual figures. www.delmarva now.com/story/news/2019/10/23/beach-nourishment-scheduled-rehoboth-dewey-beaches/406 3554002/

22 Multi-Hazard Mitigation Council (2019). Natural Hazard Mitigation Saves: 2019 Report. Principal Investigator Porter, K.; Co-Principal Investigators Dash, N., Huyck, C., Santos, J., Scawthorn, C.; Investigators: Eguchi, M., Eguchi, R., Ghosh., S., Isteita, M., Mickey, K., Rashed, T., Reeder, A.; Schneider, P.; and Yuan, J., Directors, MMC. Investigator Intern: Cohen-Porter, A. National Institute of Building Sciences. Washington, DC. www.nibs.org

23 Galveston's Bulwark against the Sea: History of the Galveston Seawall, October 1981, US Army Corp or Engineers www.swg.usace.army.mil/Portals/26/docs/PAO/GalvestonBulwarkAgainstthe Sea.pdf

24 Army Corps Releases Final $29 Billion 'Ike Dike' Study for Congressional Approval, Katie Watkins, September 10, 2021, Houston Public Media. www.houstonpublicmedia.org/articles/ news/energy-environment/2021/09/10/408118/army-corps-releases-final-29-billion-ike-dike-study-for-congressional-approval/

Part 2

Project Management Foundations and Planning for a Course of Action

Chapter 5

Developing the Project Scope

> We all know the old cliché in emergency management that "the value is not in the plan, but in the planning process". I have experienced a lot of planning processes that, at the time, really educated and trained each other, with multiple stakeholders in a room, about everyone's different capabilities, authorities, responsibilities, capacities...project management creates a life-cycle for that planning effort. (We need to be) engaging and restimulating a shared learning experience, so that knowledge is in our head.
>
> *Greg Brunelle, Assistant Secretary – Security & Emergency Management,*
> *Mass Department of Transportation & MBTA; Former Director of NY State*
> *Office of Emergency Management, New York State Division of Homeland Security and*
> *Emergency Services. Interview in 2017.*

As this quote so aptly puts it, the whole purpose of well-designed and managed planning processes is not to end up with fancy-looking charts, diagrams, spreadsheets, and reports but to engage stakeholders, including customers and developers, collaborate and coordinate our work, so we end up with outcomes that further the emergency management and public safety mission. In conversations with numerous emergency managers who have managed projects across the country and world, developing and managing the scope of a project is absolutely critical, and this is true of projects in general.

The scope of work, which includes a scope document or statement of work (SOW) and a work breakdown structure (WBS), lays the foundation of any plan. In this chapter we will:

- define the project stakeholders;
- assess stakeholders' expectations, define their needs and priorities, and begin to translate these into requirements (covered in greater detail in quality management);
- describe the scope statement and key elements;
- explain the WBS and how it is developed, and its relationship to the schedule, budget, and other project plan elements.

A principal question I want to answer at this point is: who does this work of developing the scope and other plans? Primary responsibility falls on the project manager, but project planning is a team sport, so the team and other key stakeholder need to be involved in project planning. Deciding who needs to be involved when and to what extent is the science and art of project management we will explore in the following chapters.

DOI: 10.1201/9781003201557-7

STAKEHOLDER AND COMMUNICATION MANAGEMENT

Simply put, a stakeholder is a person or group of people who have a stake in the project. This means they have an interest in it and may be impacted by or have an impact on the project. They can also expect to get something from the project, a set of benefits or meet specific needs. Stakeholders, depending on how much influence and interest they may have on the project, can be critical to a project's success. Now let's start with identifying who project stakeholders are.

IDENTIFYING STAKEHOLDERS

When it comes to emergency management and public safety, there is quite a range of stakeholders:

- Members of the public or residents in a given jurisdiction
- Agency leadership
- Agency staff
- Project team members
- Subject matter experts (SMEs)
- End-users (when it comes to adoption of new systems or technology)
- Grant funding agencies (supranational, federal, state, local, tribal, territorial, etc.; UN, FEMA, etc.)
- Businesses and the general business community
- Professional organizations or standards-based organizations (IAEM, NEMA, DRII, etc.)
- Vendors/contractors
- Private voluntary/community/ faith-based organizations (PVOs, CBOs, FBOs)
- Non-governmental organizations (NGOs; most often used internationally)
- Other government agencies within and outside, neighboring the jurisdiction; or at larger jurisdictional units, county, parish, provincial, state, national
- Mass media

The list above and diagram in Figure 5.1 describe a wide range of stakeholder categories, but not all of the stakeholder categories or examples given above would be appropriate for every project. It is important to consider them carefully based on each individual project, even if the project may be similar to a project performed before. One simple method for identifying stakeholders is to brainstorm them on the team, including the project champion or client team when possible (see brainstorming methods in quality management). The stakeholder identification process can be an iterative process, one that takes a first pass based on a draft project proposal, and then takes a second pass once the scope statement is more fully developed. You may discover, especially as you widen your scope, that more stakeholder groups are taken into account.

ANALYZING STAKEHOLDERS

Early in the life of a project, we often hear the need to get "buy-in" from our stakeholders, in order to get their input or enlist their cooperation in the project. However, not all project stakeholders have the same level of influence, nor interest in the project. With this in

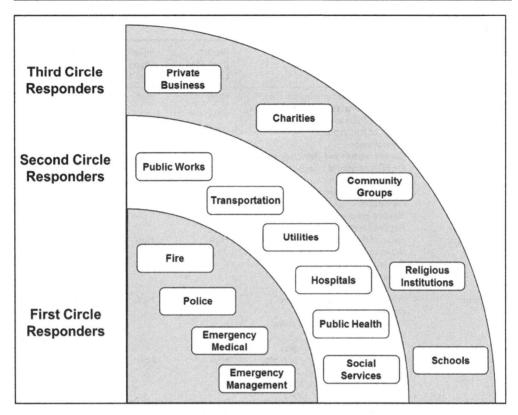

Figure 5.1 Centrality of emergency response to organizational mission. (From Nicholas B. Hambridge, Arnold M. Howitt, and David W. Giles. "Coordination in Crises: Implementation of the National Incident Management System by Surface Transportation Agencies." Homeland Security Affairs 13, Article 3 (March 2017). www.hsaj.org/articles/13773).

mind, the next step to take after identifying project stakeholders is to analyze them for three key criteria: influence, interest, and requirements. Influence on the project may translate into power that they may have over key success factors, such as controlling funding, the timeline, and scope, or who works on the project, both team members and other stakeholders.

Requirements are what stakeholders are looking to get from the project, as we described above. They may be benefits, needs to fill, or performance expectations from the outcome. We will delve into the process and methods used for gathering and defining quality requirements in Chapter 7. For now, one of the simplest methods for defining requirements is a simple interview or survey with stakeholders.

These criteria can be measured using the following stakeholder analysis matrix, provided in Table 5.1 on a qualitative basis as follows, using an example of a new program to introduce unmanned aerial vehicles (UAVs, also known as drones) at a police department:

It is important to note that this is a brief analysis of *project* requirements and not the *end-product requirements* of the project. A more in-depth analysis of the end-user and product requirements is typically done as part of the quality management process covered in Chapter 7.

Table 5.1 Stakeholder Analysis Matrix

Stakeholders	Requirements	Interest (High-Med.-Low)	Influence on Project (High-Med.-Low)	Rating (3[High]-1 [Low])
Project sponsor (Champion; Police Chief)	• Capability to find missing persons,[11] fleeing criminal suspects, and assess crime and disaster sites and damage • Image of innovation for the department • Elected leaders and public recognition for PD's additional search capability • Ease of implementation • Adherence to governing laws and regulations • Meeting granting agency's requirements (see below)	High	High	3
Project manager	• Successful project and recognition for the outcome • Deliver benefits to the police department and public • Compensation for the project	High	High	3
Project team	• Professional satisfaction of the project work • Delivering benefits to the client (police dept. and public) • Compensation for the project	High	Med.	2.5
Police officers	• Capability to find missing persons, fleeing criminal suspects, and assess crime and disaster sites and damage • Increase general public safety • Learning opportunity • Access to latest technology and an additional tool	High	Low	1.5
Vendors	• Commercial opportunity • Compensation • Client referrals/testimonial • Clear and executable contract scope	High	Med.	2.5
Grant funding agency (state or DHS)	• Benefits of the overall grant program achieved • Grant and reporting requirements (financial, demonstrated outcomes) met • Recognition (political, public) for the demonstrated program outcomes	High	Med.	2.5
SMEs	• Satisfaction of providing expertise, teamwork, and learning • Successful project outcome • Compensation (if paid; sometimes voluntary)	High	Low	1.5
Members of the public	• Increased public safety • Not abused by the PD for undue public surveillance (in some US jurisdictions governed by law[12])	Med.	Low	1
Legal advocacy and civil liberties agencies or organizations (e.g. US DOJ, ACLU)	• Not abused by the PD for undue public surveillance and use does not violate civil liberties	High	Low	1.5

Now let's look at how the stakeholder analysis matrix is used. It provides a method to prioritize stakeholders, begin developing our communications plan with them, and manage their expectations. Using the example above, we can prioritize the Police Chief over members of the public. While the project manager and team will not just ignore the public, they need to pay greater attention to the Police Chief's requirements, as the Chief has the authority to support or withdraw that support for the project if they go unmet.

These levels of influence may change over time during the project. Again, using the example above, legal organizations such as the American Civil Liberties Union (ACLU) may have little influence at project initiation, but that influence may grow as the project moves into a pilot testing stage where drones may be first introduced into public use on a limited basis, which may present an opportunity for a lawsuit to be introduced. Again, as with many aspects of project management methods, stakeholder analysis is not a static process and warrants periodic reexamination. Additionally, stakeholder communication does always take place in complete isolation between different groups, so members of the public may stoke legal advocacy organizations into action and vice versa. As mentioned above, another way in which the stakeholder matrix may be used is to actively manage expectations through communications, which we will cover next.

STAKEHOLDER COMMUNICATION NEEDS AND MANAGING PROJECT REQUIREMENTS

It is crucial to engage key stakeholders in developing the scope of work, incorporate their input, and keep them engaged in the project development process, for example developing an exercise or training. Now what does that mean, and how does that look? It means making that a part of your communications plan, and as with any communications plan, it starts with the stakeholders and defining:

- What information they need, which is related to their requirements;
- Their preferred communication methods, sometime referred to as channels or mode of communication; and
- Frequency of messaging.

We will get into communications planning and management in further depth in Chapter 7, so will hold for now with this. What is important at this point for scope development is that we have a better understanding of who the stakeholders are, their level of importance to the project, and what basic requirements they expect from the project.

SCOPE STATEMENT

A scope statement or what the Project Management Institute refers to as the Project Charter, sometimes called a preliminary scope statement, is a document that authorizes the project and provides the project manager the authority to move forward with the planning process. While there are varied formats in use by different types of organizations and project managers, a scope statement generally comprises the following elements:

- Project name
- Project key stakeholders: client(s), project manager, internal sponsor (champion), and date
- Goal
- Business case
- Deliverables, phase, or major objectives
- Critical success measures
- Constraints, assumptions, risks, and exclusions
- Major milestones, deadline, and estimated budget

Let's explore each of these further using the following example based on the project example (Table 5.2), the development of an online course for a Disaster Recovery Center.[1]

Table 5.2 Project Scope Document

Project Name	Disaster Recovery Center (DRC) Online Course9
Project Client Team	Rose Menendez, Assistant Commissioner, Social Services Agency (HSSA), Hazard City[10]
	Noel Miller, Director of Social Services Support, Social Services Agency
	George Sidwell, Director of Training, Office of Emergency Management (OEM), Hazard City
Project Internal Sponsors (Champions)	Mary Baker, Director for Human Services, OEM Hazard City
Project Manager	Carla Washington, Project Manager
Effective Date	January 15, 20XX

Goal: Develop a 2-h online course within four months for Hazard City employees who will staff the Disaster Recovery Center (DRC) that addresses the functions, organization, roles, and services offered by the center, including how to stand up, operate, and demobilize, as part of the post-disaster response.

Description and Business Case:

In response to a major disaster, the Hazard City Government may open one or more DRCs to provide a central location to serve post-disaster recovery needs for those who have been impacted by the disaster. This may include social services, relief assistance, and information from local, state, and Federal agencies and voluntary service organizations. At the moment, an in-person course is provided and takes one day of training and is delivered to 1,000 Hazard City workers on a bi-annual basis. The HSSA and Hazard City Office of Emergency Management (OEM) came together and determined that to save costs in delivering this course and to provide for a just-in-time (JIT) training option, an online delivery option was a practical alternative to deliver the course.

Deliverables:

1. Project Plan (scope, schedule, budget)
2. DRC Online Course (120 min)
   - Course design document
   - Storyboards for course
   - Alpha version (slides, script, and images for course)
   - Beta course version for pilot testing
   - Pilot test

- Final (Gamma) version, with electronic files, course certificate, SCORM and Section 508 compliant[2]
3. Testing plan
4. Course evaluation survey

Critical Success Measures:

- Evaluations achieve a score of 4 or higher from the final course on a Likert scale[3] (5-point) on key metrics: understanding the functions, roles, processes, procedures of the DRC and confidence in working on a DRC.
- Course meets SCORM and Section 508 requirements.

Constraints:

- Grant funding deadline (end of fiscal year in June, 20XX).
- Fixed price contract (budget) $120,000.
- Section 508 compliance.
- Review and approval by the Hazard City Legal Dept.

Assumptions:

- Qualified eLearning course developers.
- Stable content-approved for the DRC Plan (including job action sheets, software system) and classroom-based course.
- Target Hazard City employees available for pilot tests.
- Access the Hazard City learning management system.
- Development reviews are conducted according to schedule and delivered on a timely basis.

Risks:

- Changes to content-approved DRC plan and course.
- Emergency incidents (responses to disaster events) may take HSSA and H-OEM staff away from input into course development and reviews.
- Technical issues with the online course in the Hazard City LMS.
- Does not meet Section 508 compliance.
- Does not pass legal review.

Exclusions:

This project does not include

- User support.
- Development of additional job aids (checklists, field operations guide, etc.)
- Any updates or edits to the classroom-based course (PPT slides, instructor guide, or participant manual).

Milestone Dates:

- Feb. 1 – Design document approved
- March 15 – Storyboards for course approved
- April 1 – Alpha version
- April 15 – Beta course version for pilot testing
- April 30 – Pilot test
- Final (Gamma) confirmed complete May 15

Estimated Budget: $120,000

Let's walk through each of the scope statement elements in the example above.

GOAL

A project begins with understanding what the intended goal of the project is from a high level. In order to make this clear, the goal needs to be described in a SMART format: specific, measurable, achievable, realistic, and time bound. For that matter, objectives related to the goal also need to be described in a SMART format. SMART goals and objectives meet the following criteria:

- *Specific* means that the goal needs describe a specific outcome and address the five Ws: who, what, where (*when* is answered by the T in time-bound; and *why* is answered by the business case).
- *Measurable* means that the goal has a quantitative or objective metric by which it can be declared to satisfy the end result.
- *Achievable* means that it can realistically be achieved with the capabilities, including resources, and time deadline.
- *Relevant* means that it is aligned with the organization's objectives and provides a meaningful set of benefits or meets the performance requirements of the end-users or community.
- *Time-bound* means that there is a clear deadline that the project needs to meet.

In the many years that I have been working in project management, stating the goal using a SMART format is an effective method for starting off with a well-defined scope statement. A poorly stated goal is like a loose cornerstone in the foundation of a project plan and only opens the possibility of scope creep. Scope creep is when additional requirements creep into the project as key stakeholders, who have a great degree of power, reinterpret the project goal, and possibly define unrealistic expectations, with insufficient resources and unreasonable deadlines, outcomes, or metrics. The result of this is usually costly rework, uncompensated hours of work, and a demoralized team. In contrast, a clearly stated SMART goal provides a clear path forward for the team and great comfort being able to refer to the goal and communicate that to different stakeholders.

BUSINESS CASE

We need to understand the business case or, plainly, the justification for the project. This clearly answers the questions:

- Why are we undertaking this project?
- Why are investing valuable resources in this project?
- What are the potential project benefits?

In the example given above for the online DRC Online Course Project, there are two key reasons for this project: saving staff time for training and providing a JIT training course as an option. A quick "Back of the Envelope" analysis of staff time and course delivery would yield the following analysis:

8 h of training × 1,000 staff = 8,000 h
50 training sessions (assuming 20 staff/course) × 8 h/course instruction = 400 h

If we use a factor for hourly costs for staff, including any fringe benefits and overhead costs (which we will get into Chapter 6), we can see a clear benefit in savings from the online course that outweighs the cost of the classroom course delivery. We can do the same, using a cost per hour to calculate the training delivery. If a majority of these costs are eliminated by offering the online course option, it justifies the project. Then there is the intangible and unmeasurable benefit of the online course for JIT training. It is just not feasible to deliver a classroom course as a JIT training to a broad range of staff efficiently, consistently, and flexibly is an emergency deployment.

DELIVERABLES, PHASES/OBJECTIVES

From the goal we can break this down into deliverables, phases/objectives in the project. When it comes to the next level in which we break down the goal, it can be thought of in two ways: as a culmination of all of the deliverables or objectives or the pinnacle achievement of the project phases. According to PMBOK© (PMI), "a deliverable is a verifiable product, result, or measurable service performance capability" that constitutes the overall project goal. A deliverable can be a product component, a document, or a piece of software. In principle, a deliverable should be something that you can see, touch, or experience and basically something that can be "delivered" to the client or end-user that they can see and work with as a complete result. Referencing the example above, the goal is to develop a DRC online course and the deliverables that constitute that goal are:

- Course design document
- Storyboards for course
- Alpha version (slides, script, and images for course)
- Beta course version for pilot testing

- Pilot test
- Final (Gamma; SCORM and Section 508 compliant) version, with electronic files and course certificate

The other way of breaking down the goal is by phase/objective. A phase is a series of activities that result in a deliverable. Phases are usually sequential in order, and both deliverables and phases are not mutually exclusive when breaking down the goal. For example, if we defined them for the DRC online course development project, it might look like the following:

- Design Phase
- Phase 1 – Storyboards
- Phase 2 – Alpha
- Phase 3 – Beta
- Phase 4 – Final Development

An objective is similar to a phase, the major difference being it is defined as a strategic point to be achieved along a path to an end point. These terms are sometimes used interchangeably when it comes to project management. However, in emergency management, the distinction between a phase and an objective is much clearer and more critical. When a disaster incident occurs, objectives are established by the Incident Command on down through the chain of command and drive the emergency response.

My experience has been that deliverables are easier to define than starting with phases. When it comes to developing the WBS, which defines the tasks (sometimes called work packages), it is sometimes difficult to know when a phase begins and ends, and knowing whether the tasks roll up to complete a phase, making it 100% complete. We will explore this in greater detail in the WBS section later in this chapter.

Most basic projects in emergency management and public safety can be categorized into four common areas: plan development and job aids, training development (which may include job aids), exercise development (all types), and adopting new technology and systems. Each of these projects also has a set of deliverables common to it. Some examples of these deliverables specific to these projects will be provided as we go through the following chapters.

CRITICAL SUCCESS MEASURES

Critical success measures are measures which demonstrate that the project is a critical success. Beyond meeting the deadline and coming in under budget, these are the indicators that show whether the goal has clearly been met, using criteria that allows for some level of objective measure, rather than relying on subjective opinion, that the project will provide the expected outcome. Defining measures may be difficult for some projects, such as the online course example here, so in this case, using an evaluation survey, we can then turn end-user feedback into measurable outcomes. The standards-based criteria, SCORM and Section 508 compliance, also provide objective measures that are critical for it to meet. We will relate how quality management relates to critical success measures when we cover quality in Chapter 7.

There are some projects in emergency management, for example exercises, which are designed to test response plans or systems and point out organizational performance gaps, but which may be carried out in a way that they are successful in doing so. In essence, the emergency response plan might fail to perform well in the exercise, but the project to exercise the plan may succeed in pointing that out.

CARE: CONSTRAINTS, ASSUMPTIONS, RISKS, AND EXCLUSIONS

After defining the goal, deliverables or phases, and the critical success measures, we go onto our next major step, which is to ensure these are handled with *CARE*; that means *Constraints*, *Assumptions*, *Risks*, and *Exclusions*. Why do I use the term CARE here and why is this important? The CARE items in a scope statement act like a set of preliminary guard rails for the project team as they begin to further elaborate the project plan. Referring to Figure 5.2, we can imagine the path taken on a project as a road to a pinnacle of a mountain, with the top as the goal. The winding path is bound on each side by the CARE terms to ensure the team stays on the road and avoids potential pitfalls. These terms are explained in greater detail below.

CONSTRAINTS

Constraints are limitations on the project, describing specific factors that limit access to resources, certain actions, or design options. Almost always, time and budget are the paramount constraints, but there may also be specific constraints such as legal or regulatory requirements. There may also be organizational limitations. A simple example of a common constraint for emergency plans, facilities, or services is the need to meet ADA requirements or what are called Disabilities Access and Functional Needs (DAFN).[4] Some of these constraints may be jurisdictional, such as ensuring that disaster assistance is

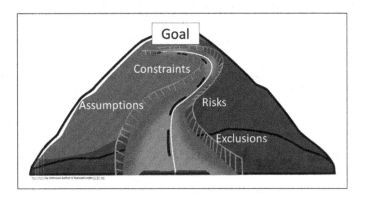

Figure 5.2 Constraints, assumptions, risks, and exclusions on the path to the goal.

administered in an equitable manner, using such indices as the Social Vulnerability Index (SVI)[5] for resource allocation decisions.

ASSUMPTIONS

Assumptions are those items available or conditions present that we assume to be there for the project. On some projects we may make assumptions around team members, having stable requirements, or that the business case does not change. In our example above, we make an assumption that the reviews are conducted in a timely manner. However, we need to question our assumptions on projects, so if we are unsure about them, they may become risks, which we address next.

RISKS

We have discussed risks at length in our previous module when it came to how, and why, emergency or public safety projects are undertaken, but here we are looking at them from a project planning perspective, considering risks more specifically as to how they might impact a project. At this stage of planning, we don't have to elaborate all project risks but just the major ones the project manager or team may be aware of. A review of the preliminary parts of the scope statement, deliverables, success measures, constraints and assumptions may yield some ideas for risks. Referring to the example again, we can identify a number of risks that relate to critical factors for project development: content, compliance, technical issues, etc. Again, a reminder here, the focus should be on risk events, with a mind on how can we analyze and manage these risks. We will explore project risk management in further depth in Chapter 7.

EXCLUSIONS

Finally based on our understanding of the three former categories, we may want to exclude certain options or actions, features or items that we may deem outside of our budget, not feasible given the time or public demands, or that the client may deem to be unnecessary. It is just as important on projects to be clear about what you plan to develop and what you will not develop, what is outside the scope. This will avoid misunderstandings later on during the development process. In our example, we identified specific items, such as the development of job aids or updates to the classroom courses (on which the eLearning course is based) as excluded, as well as user support which, on its own may not be a capability of the project team or achievable within the timeline.

SCHEDULE, MILESTONES, AND BUDGET

We finish this initial scoping activity by defining the high-level schedule (also may be called a timeline) by identifying major milestones or phases, with dates or time to complete, and an estimated budget. We will address scheduling and defining related terms in Chapter 7, but for now let's make sure to define a milestone: a significant point or event in the project, often related to the end of a phase or objective. Milestones are often used as a review or stage gate, sometimes for high-level approvals before moving on to the next stage of a project. Milestones are sometimes provided by the client and are driven by a deadline or proposed start time.

The budget is an estimate of the cost of the project based on our current understanding of the work. Similar projects completed in the past may provide a rough estimate for the project budget and schedule. A schedule or budget at this stage is often notional, providing a general idea. A more detailed schedule and budget are developed once we define the detailed WBS and further elaborate the project plan. Remember, developing the project plan and managing it on an ongoing basis is an iterative process.

FINAL NOTES ON THE SCOPE STATEMENT

The scope statement is an early document that gives a quick snapshot of what the project is designed to do. A requirements document may be a part of this, and we will get to this when we discuss project quality management. This early scope document may seem to be a minor document, but it has major implications as it can serve as a means to get major stakeholders on the same page and confirm their expectations, in order to avoid confusion or misinterpretation. My recommendation is during a project kick-off meeting to read the scope statement aloud and pause at each step to ask if there are questions. This helps to avoid any lack of understanding or commitment to what was agreed to.

WORK BREAKDOWN STRUCTURE

The WBS is the next level of detail for scope development beyond the scope statement. A WBS is defined in the PMBOK© (PMI) as: "a hierarchical decomposition of the total scope of work to be carried out by the project team to accomplish the project objectives and create the required deliverables". The WBS forms the core of the project plan, and the schedule, budget, and other project plan components are based on it, as illustrated in Figure 5.3. A well-defined WBS helps to develop an effective project plan. However, always bear in mind that as other elements of the plan are developed, you will need to periodically look at whether additional tasks or work may need to be added.

A WBS may be developed using a hierarchical chart form or an outline with indented items, an outline version is provided here. Let's use a different example this time to demonstrate a WBS, the development of a Table-Top Exercise[6] (TTX) for assessing a Business Continuity Plan for an organization (for government, termed a COOP Plan). I am including two versions of this WBS: an organization chart structure, in Figure 5.4, and an outline version. The latter is useful when transposing this into a software program and for creating a budget in a spreadsheet program.

Figure 5.3 WBS as core of the other project planning elements.

WBS Outline format for the TTX Exercise Project

0.0 Develop and conduct a TTX to assess the Business Continuity Plan for ABC company within three months.
1.0 Project Plan (Scope, Schedule, Budget, Quality, Risk, HR, Stakeholder/Communications Procurement)
    1.1 Develop plan elements.
    1.2 Review plan elements with team, client, and key stakeholders.
    1.3 Approve plan and manage project in alignment with plan.
2.0 Planning Meeting
    2.1 Identify and schedule planning meetings with key stakeholders
    2.2 Develop agenda and gather planning materials
    2.3 Conduct planning meeting with key stakeholders and follow up on next steps with TTX development
3.0 Situation Manual
    3.1 Develop the draft Situation Manual: goal, objectives, key capabilities to assess, scenario, questions, etc.
    3.2 Review draft Situation Manual and collect edits
    3.3 Develop final version of Situation Manual and confirm complete
4.0 Exercise Evaluation Guide (EEG) and Evaluators
    4.1 Develop the draft EEG
    4.2 Review draft EEG and collect edits
    4.3 Develop final version of EEG
5.0 Presentation
    5.1 Design presentation
    5.2 Develop draft presentation
    5.3 Review presentation and collect edits
    5.4 Develop final version of presentation and confirm complete

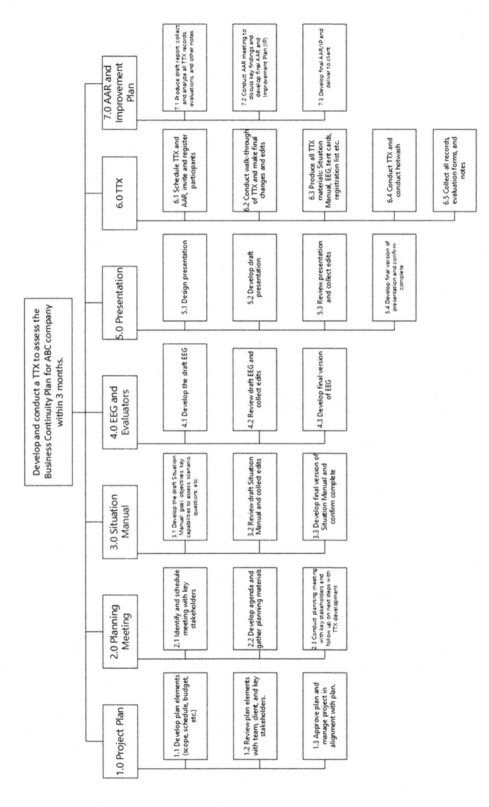

Figure 5.4 Work breakdown structure Org. Chart format for the TTX Exercise Project.

6.0   TTX
    6.1 Schedule TTX and AAR, invite and register participants
    6.2 Conduct walk-through of TTX and make final changes and edits
    6.3 Produce all TTX materials: Situation Manual, EEG, tent cards, registration list, etc.
    6.4 Conduct TTX and conduct hotwash
    6.5 Collect all records, evaluation forms, and notes
7.0   After-Action Review (AAR) and Improvement Plan
    7.1 Produce draft report: collect and analyze all TTX records, evaluations, and other notes.
    7.2 Conduct AAR meeting to discuss key findings and develop final AAR and Improvement Plan (IP)
    7.3 Develop final AAR/IP and deliver to client

WBS STRUCTURE AND WORDING FORMAT

The scope statement provides a starting point for the WBS, and there are a number of steps and rules to follow when developing a solid WBS, both in its structure and the word format. As the definition states, the development of a WBS is a decomposition of the work to be performed, and we do this using the following major steps starting with the overarching goal:

1. Put the goal at the very top and begin with the major deliverables (preferred method) or project phases from the scope statement. Deliverables are described in a noun format, so it is clear that this is a higher level outcome and not a lower level work activity.
2. Define tasks at the next level, in other words the work to perform to build toward the deliverables/phases. Tasks are sometimes referred to as "work packages" in the nomenclature of project management; for our purposes we will refer to these as tasks and also to avoid confusion with objectives or phases. Tasks use a verb-noun format, making them action oriented to clearly indicate the work being performed and that there is a defined outcome. From the WBS example above, task *7.3 Develop final AAR/ IP and deliver to client* makes it clear what work is being performed and completed for whom, using some of the SMART criteria we discussed earlier.
3. Indent the next level down to the right to indicate that it is a subordinate item and rolls up into the level above.
4. Number each of the elements, using a hierarchy of sequential numbers followed by a decimal to denote a WBS level. Numbers are used, so we can easily trace the work as we begin to develop other plan elements (schedule, budget, etc.) and when tracking the progress of work. In the TTX example, items at the X.0 level are deliverables, and the next level down is what are called tasks and numbered X.1, X.2, etc.

If using a type of diagram such as an organization chart, the numbering can use the same method, and subordinate items can be indicated by using lines to show the subordinate connection of tasks to the deliverables, and deliverables to the goal.

What method should be used to develop the tasks? A team (not just the project manager) may use a brainstorming method to identify tasks based on the deliverables. Another way to start is to use existing standards such as EMAP, HSEEP, and NFPA 1600 which provide an outline or methods to develop plans, training, exercises, etc. You can also borrow existing plans from other jurisdictions or subject matter experts, especially for new technology/systems development. One word of caution when using previous or borrowed plans; make sure to carefully review the plans to adapt them to your specific needs, organization, and working environment, identifying tasks unsuitable for your project and those that may be required.

WBS RULES – LEVEL 1, 10, AND 100%

Questions often arise as to what level work should be broken down and what is the length of a task. You can think of these using the following mnemonic device Level 1, 10, 100% which related to these rules to follow for a well-defined WBS:

- **Level:** break work down to a level at which you can reasonably estimate and manage the work
- **One:** a deliverable should be expressed as one complete unit or part, not multiple items combined into one
- **Ten days:** tasks should not be longer than ten days long or shorter than two days
- **100% rule:** each level "rolls up" to the next higher level – must be 100% complete

You will notice that there is no rule on sequence or timing for each task. That is done in the next phase of project development. Putting the tasks in logical order, while desirable, is not an absolute requirement at this point. Let's explore each of these rules and why they apply.

LEVEL

The paramount rule here is breaking work down to the level at which we can reasonably estimate and manage the work. If we develop a task that is too large, our estimates may be faulty and hamper our ability to manage that work. Taking deliverable number 3. Situation Manual from above, if we were to only have one task, 3.1 Develop Situation Manual and estimate that at three weeks for development, our estimate may have questionable accuracy, and may also be missing some key details. Let's say we assign this work to an individual, and they report back that they are 50% done at the two-week mark, we can only assume that the work may take another two weeks based on the progress, and with additional time for review, would translate into over four weeks. However, if we have broken it down as we have here and task 3.1 (with a complete a first draft) is supposed to be finished and ready for review within a week, then we have a more reasonable level to manage the work and take corrective actions.

ONE WHOLE DELIVERABLE

When each deliverable is one complete unit, this allows us to ensure that we do not have unnecessary dependencies holding back the completion of a unit of work. Using the example above, if we had one deliverable titled "Documentation" which included the Situation Manual, EEG, and Presentation, we could not consider it complete until all items were complete, we may have challenges estimating the work and figuring out when it might be complete. Separating each of these deliverables makes managing them easier.

TEN DAYS

Until we start assigning tasks to a calendar, using a Gantt chart or equivalent, we are referring to work days in a project, not calendar days, such as weekends, holidays, or other non-working time (explored further in Chapter 6). Ten days is typically equal to two work weeks and is a reasonable time horizon to work within. My general experience has been that allowing a task to go more than two weeks without checking on its status often leads to unnecessary delays. On the other hand, we do not want to estimate a task at too limited a time frame, as having too many small tasks, fills the schedule and makes it too detailed and unwieldy to manage. We do not want to be managing a section-by-section development of the Situation Manual in the example above. A minimum of two days or roughly 10 h is a general rule to apply to tasks that are identified.

100% RULE

The final rule here is that 100% of the subordinate items below must roll up to complete the item above; tasks below a deliverable must role up to complete the deliverable, and the deliverables must roll up to complete the goal. This requires professional expertise in a given field and teamwork to make sure that a deliverable is broken down into the key tasks. This includes reviews and approvals and external production (of hard copies) and proofing, which sometimes go missing from a WBS but are critical to ensure completeness.

SCOPE MANAGEMENT

As we will cover in our next chapter, the scope of work, as constituted by the scope statement and WBS, provides a foundation to develop the schedule, budget, quality, risk, HR, stakeholder/communications, and procurement plans. Once we develop each of those plans, a project manager and the team should go back to the scope and look at what might need to be updated. This process is called project integration and aligns each project element (scope, schedule, budget, quality plans, etc.), so they all work together, instead of as disparate units. Once the project plan is approved, the WBS is actively managed, usually as a part of a Gantt chart or similar scheduling tool, but it is worth maintaining the WBS in one place to make sure it is current, linking details (checklists, guidelines, etc.) to them as necessary.

Any changes introduced to the scope, including the deliverables or requirements that have been identified (usually done as part of the quality management process), need to undergo a thorough review called a change control process review. This includes a look at how it impacts all of the other project elements, most importantly the cost and schedule. Once a change has undergone a change control process and impacts determined, then it is either approved or denied, and this outcome documented. If it is approved, then that approval document is received, and the change is then made to the overall plan, including each element, and new plan (new scope document, schedule, budget, etc.) is approved for implementation.

GENERAL PRODUCT DEVELOPMENT GUIDANCE

If you are developing a project from scratch, without any previous project plans or concept of the work tasks, then you may want to brainstorm the tasks as a team, using the deliverables as your starting point.

My experience is that a good number of the design and development phases can be defined as follows:

1. Identify key requirements (with the client/end-users).
2. Create the design document (this may be an outline, blueprint, basic framework or "wire frame"; term used in the world of user experience [UX] design).
3. Review design (by client, SME, team, others).
4. Develop prototype (Alpha version).
5. Review and revise.
6. Develop working version (Beta).
7. Pilot and revise.
8. Complete final revisions and deliver final version (Gamma).

This is the classic "waterfall" development process: Design→ Alpha→ Beta→ Gamma versions (Final; other developers use terms such as Bronze, Silver, Gold; or Straw Man [Design], Woodman, Tinman, Ironman; choose your motif). Alpha is a working version that can be reviewed by a client or end-users, and the Beta version is one ready for pilot testing. All changes flow from review of the previous version until you get a final approved version of the product, e.g. system, training, and emergency plan.

FINAL THOUGHTS ON SCOPE MANAGEMENT

A majority of project failures are attributed to poor scope definition,[7] often what is cited as scope creep.[8] Scope creep is essentially the addition of functions, features, or requirements to the project that were not a part of the original scope of work, for which there was no agreed to change control, and no subsequent change in schedule and budget. What are the consequences? Running over budget, beyond the deadline, and not meeting the requirements the project originally set out to meet. This often does not address the human toll in additional hours of extra work or rework, frustration, and reduced morale.

Why does scope creep happen? In my years of managing projects and researching them, some key reasons:

- Poorly defined scope documents, including requirements.
- Not setting clear boundaries for clients and the implications of changes (similar to that above).
- Beginning a design and development process before a thorough needs analysis and requirements definition has taken place (we will address this under quality management, but this closely relates to scope).

Ultimately, most of these reasons boil down to introducing changes to the scope without any careful consideration under a change control process. A clear mitigation measure to reduce the likelihood of scope creep is to make sure you have a tight scope development and management process, including a change control process in place. We will cover project plan and execution management later in the book.

NOTES

1 A Disaster Recovery Center is a central location where disaster survivors can seek assistance from governmental agencies and private voluntary organizations as a "one stop shop" to provide for all their needs. This can be a used by federal, state, and local government agency representatives (emergency management and social service agencies) and service organizations (United Way, American Red Cross, Feeding America, Salvation Army, Southern Baptist Convention Disaster Relief, etc.). While this term is used by FEMA (US DHS), this is used here purely as an example. This and the following examples are fictional, but based loosely on examples and better practices. Any resemblance to actual persons, living or dead, places, or actual events is purely coincidental.

2 SCORM is an eLearning (online) course compatibility standard commonly used for learning management systems, see: https://scorm.com/. Section 508 compliance is the online standard equivalent of ADA accessibility, see www.section508.gov/

3 A Likert scale is commonly used to measure attitudes, knowledge, perceptions, values, and behavioral changes. A Likert-type scale involves a series of statements that respondents may choose from in order to rate their responses to evaluative questions (Vogt, 1999). See: https://mwcc.edu/wp-content/uploads/2020/09/Likert-Scale-Response-Options_MWCC.pdf

4 According to FEMA, "individuals in need of additional response assistance may include those who have disabilities; have limited English proficiency or are non-English speaking; or are transportation disadvantaged". See: www.disasterassistance.gov/information/disabilities-access-and-functional-needs/online-resources

5 Social vulnerability are factors, including poverty, lack of access to transportation, and crowded housing that may weaken a community's ability to prevent human suffering and financial loss in a disaster. The CDC/ATSDR SVI uses US Census data to determine the social vulnerability of every census tract. www.atsdr.cdc.gov/placeandhealth/svi/index.html

6 A Table-Top Exercise (TTX) is a discussion-based exercise that is often used to help assess a plan or as the basis for identifying gaps in existing capabilities to define critical planning elements, resources, systems, training, or new plans that need to be developed.

7 Dumont, P. R., Gibson, G. E., & Fish, J. R. (1997). Scope management using project definition rating index. *Journal of Management in Engineering*, 13(5), 54–60. https://doi.org/10.1061/(ASCE)0742-597X(1997)13:5(54)

8 Larson, R. & Larson, E. (2009). Top five causes of scope creep … and what to do about them. Paper presented at PMI® Global Congress 2009—North America, Orlando, FL. Newtown Square, PA: Project Management Institute.

9 This example is used only for illustrative purposes and is not intended to represent actual figures. Any resemblance to an actual project is purely coincidental.

10 Hazard City is used here as part of a fictional location or provide a context to illustrate certain organizational and community dynamics.

11 Police Executive Research Forum. 2020. *Drones: A Report on the Use of Drones by Public Safety Agencies—and a Wake-Up Call about the Threat of Malicious Drone Attacks.* Washington, DC: Office of Community Oriented Policing.

Service Drones have been most commonly used for silver or amber alerts in specifically defined zones, sometimes outfitted with infrared cameras to detect heat signatures.

12 **How to Regulate Police Use of Drones** September 24, 2020 Faine Greenwood. The Brookings Institution – Tech Stream. See: www.brookings.edu/techstream/how-to-regulate-police-use-of-drones/

Chapter 6

Developing the Project Schedule

> If you don't choose to do it in leadership time up front, you do it in crisis management time down the road.
>
> *Stephen Covey, Management Consultant and Author*

This is a good reflection on projects carried out in an emergency or public safety organization. As we alluded to previously, the exigencies of the immediate work in an emergency or public safety agency often take precedence over project work, which often is an additional assignment to busy schedules. This is even more the case when working cross-functionally with different departments, agencies, or external partners, be they other governments, private, community, or other groups, many of which have competing priorities or their own constraints to deal with.

When people talk of project plans, one of the first images that comes to mind is the ubiquitous Gantt[1] chart. Even though the Gantt chart is an effective visual tool to communicate a project schedule, there are many key methods and tools that precede its development and use. It is also important to point out that the schedule is only one part of a project plan.

PURPOSE OF A PROJECT SCHEDULE

A project schedule serves several purposes, as a planning tool for helping to schedule and coordinate work with different resources (people, equipment, etc.), establishing start and finish times, meetings (reviews, coordination, updates), and milestone events. It provides a snapshot of progress for tracking work performance and reporting that helps to forecast future progress, indicating whether the project is on or off track for the completion date. A project should be a dynamic tool that is developed and updated on a regular basis with the project and client team, as well as key vendors, and to engage them throughout the project as opposed to a static timeline that soon becomes out of date.

MAJOR STEPS IN DEVELOPING A PROJECT SCHEDULE

Developing a realistic schedule is a progressive process that builds on each preceding step. These steps are as follows:

DOI: 10.1201/9781003201557-8

1. Create a network diagram using the project tasks from the WBS.
2. Estimate the time required to perform the work for each task.
3. Elaborate the project schedule using calendar dates and constraints and developing a timeline, a Gantt chart or other visual calendar chart.

These steps are broken down below and into sub-steps in the schedule development process.

DEVELOPING THE NETWORK DIAGRAM

What is a network diagram? A network diagram is a graphical view of the logical sequence (sometimes referred to as a logic network) of project work (tasks/work packages). The network diagram helps to estimate the total project duration by highlighting the critical path (longest path through the network), and what slack time (the period a task can be delayed without impacting the overall duration; this is what PMI terms "float") may exist on specific tasks or segments of the project. The network diagram is an essential element to accurately develop the schedule. A quick reminder here that with any major planning effort, it is a team sport and should involve the entire project team.

Steps in developing a network diagram:

1. Identify all work tasks.
2. Establish task relationships (precedence; finish to start, start to start, finish to finish).
3. Develop the diagram from start to finish; or, as I prefer from finish to start, working backward.

(Sub) Step 1: Identify All Work Tasks

Identifying all work tasks is usually done based off of the WBS. We do not use the deliverables in the network diagram as they are essentially a summary of all the tasks under them, so it would be redundant for scheduling purposes. While some project management software does allow deliverables to be displayed and summarizes the overall duration for that portion of the schedule, we do not include them when separately building the network diagram. As stated above, for projects that are new to the organization, some teams develop tasks using a brainstorming technique to identify tasks.

It is also advisable to speak with stakeholders if there are any critical stage gate reviews, inspections, approvals, or other design and development activities that are required for the project. In turn these can be added to the WBS. In principle, any activities that accumulate any costs on the project should be accounted for. We will get into this further when we discuss cost estimating and budgeting.

Step 2: Establish Task Relationships

Table 6.1 is an example of a *Task Sequencing Chart* that illustrates how we begin to manually develop a network diagram for the tabletop exercise (TTX) project we presented

Table 6.1 Task Sequencing Chart

Task# and Description	Immediate Predecessor/ Successor	Task Relationship	Task Duration Days
Start	None	None	0
1.1 Develop plan elements (scope, schedule, budget, etc.)	Start	Finish-Start	2
1.2 Review plan elements with team, client, and key stakeholders.	1.1	Finish-Start	5
1.3 Approve plan and manage project in alignment with plan.	1.2, Finish	Finish-Start, Finish-Finish	35
2.1 Identify and schedule planning meetings with key stakeholders	1.1	Finish-Start	2
2.2 Develop agenda and gather planning materials	1.2, 2.1	Finish-Start	2
2.3 Conduct planning meeting with key stakeholders and follow up on next steps with TTX development	2.2	Finish-Start	2
3.1 Develop the draft Situation Manual: goal, objectives, key capabilities to assess, scenario, questions, etc.	2.3	Finish-Start	3
3.2 Review draft Situation Manual and collect edits	3.1	Finish-Start	5
3.3 Develop final version of Situation Manual and confirm complete	3.2	Finish-Start, Finish-Finish	2
1.1 Develop the draft EEG	3.1	Start-Start, 3 days lead	3
1.2 Review draft EEG and collect edits	4.1	Finish-Start	5
1.3 Develop final version of EEG	4.2	Finish-Start, Finish- Finish	2
5.1 Design presentation	3.1	Finish-Start	2
5.2 Develop draft presentation	4.1, 5.1	Finish-Start	2
5.3 Review presentation and collect edits	5.2	Finish-Start	3
5.4 Develop final version of presentation and confirm complete	5.3, 6.2	Finish-Start, Finish- Finish	2
6.1 Schedule TTX and AAR, invite and register participants	2.3, 6.4	Finish-Start, Finish-Finish	20
6.2 Conduct walk-through of TTX and make final changes and edits	3.3, 4.3, 5.4	Finish-Finish	2
6.3 Produce all TTX materials: Situation Manual, EEG, tent cards, registration list, etc.	6.2	Finish-Start	5
6.4 Conduct TTX and conduct hotwash	6.1, 6.3	Finish-Start	1
6.5 Collect all records, evaluation forms, and notes	6.4	Finish-Start	1
7.1 Produce draft report: collect and analyze all TTX records, evaluations, and other notes.	6.5	Finish-Start	5
7.2 Conduct AAR meeting to discuss key findings and develop final AAR and Improvement Plan (IP)	7.1	Finish-Start	1
7.3 Develop final AAR/IP and deliver to client	7.2	Finish-Start	3
Finish	7.3	Finish-Start	0

in the previous chapter. This identifies the specific tasks and relationships between these tasks. This chart also identifies the task durations.

The chart above is just one way to take the first step and begin sequencing tasks in a logical order.

The second step is to then sequence the tasks as a diagram, and this is a great team-building exercise that you can do in-person. I recommend taking large format Post-It© notes or similar materials, and a large white-board wall or butcher-block paper attached to a wall or laid on the floor, and sequence tasks in a linear fashion from either the start point or finish point (both are milestone points in the project).

TASK RELATIONSHIPS AND HARD VS. SOFT LOGIC

When developing a network diagram what we are looking to do is to establish the logical sequence of tasks, which tasks need to precede and succeed others. In order to do this, we need to understand the relationships between tasks. These task relationships are defined as follows in Table 6.2.

In Figures 6.1 and 6.2 a series of diagrams illustrate the top three relationships as they would appear in a section of a network diagram.

Tasks can have multiple relationships to one another, where one task may precede multiple tasks or vice versa.

Some task relationships have a hard logic, meaning that they must absolutely precede the other task (as given above), or they may have a soft logic which means that the relationship allows some level of flexibility or discretion. Some examples of hard logic are physical or regulatory limits: work cannot proceed until the contract has been signed; or you cannot conduct the TTX without printed copies of the Situation Manual and Exercise Evaluation Guide. An example of soft logic is when it may be desirable to complete all

Table 6.2 Task Relationships and Examples

Task-to-Task Relationship	Description	Example
Finish to Start	The predecessor task must finish before the next task may start.	Complete the design document and approve before plan development may begin.
Start to Start	The predecessor task must start before the successor task may start. This may often involve a lead time for the predecessor.	In the example above, the EEG has a start-to-start relationship with the Situation Manual, with a three-day lead time.
Finish to Finish	The predecessor task must finish before the successor task may finish. This may often involve a lag time for the successor to start.	In the example above the presentation has a finish-to-finish relationship with the walk-through, assuming that additional changes may require updating the presentation.
Start to Finish	The predecessor task must start before the successor task may finish. This may often involve a lag time for the predecessor. This is a rare relationship and can mostly be explained in the other three task relationships.	A good example of this relationship is when an old software system is phased out of use only after the new system is running smoothly. In principle this may also be stated as a finish to start with a lead for the start time so that we accommodate overlap between the two tasks.

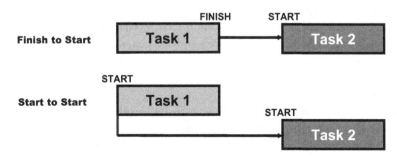

Figure 6.1 Finish-to-start and start-to-start relationships.

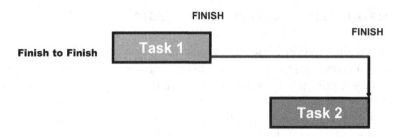

Figure 6.2 Finish-to-finish relationship.

elements of an emergency mass distribution plan for food commodities (for a post-disaster event) before developing a training course for the plan staff, a project team may still decide that a training development team can go ahead with early designs and development activities once 50% of the plan is developed vs. the full 100%. It is important to understand the difference between hard and soft logic as it plays a significant role in determining when schedules might be accelerated and what constraints there are on a schedule. It is also critical to confirm an understanding around any milestone (also called stage gate) reviews to take into consideration any hard vs. soft logic present.

An example of a completed network diagram is given in Figure 6.3. Note – this diagram only includes the task relationships, WBS#, and task duration.

Note – on the chart and diagrams above I used a Start and Finish as milestones. A milestone is a task with zero duration that is used to indicate a stage in the project schedule. Milestones can be used to define the starting, finishing, or an intermediate point for multiple tasks. I find that they are useful as a method to pull multiple tasks or work streams together when you have a number of complicated work-flows and/or a stage gate review.

A couple of common pitfalls that you may encounter when developing a diagram:

- Do not leave "orphaned tasks". These are tasks without any relationship to other tasks, with no predecessor or successor. All tasks should be part of the network diagram and have a relationship with a predecessor or successor, even if it is the Start and Finish, or other milestones.
- Sequencing all tasks in long chain of work, with finish-to-start relationships, with no parallel tasks or work streams is incorrect. Most project task networks have tasks which may be performed in parallel.
- Sequencing tasks by deliverables is a shortcut, but it violates the level and 10-day rules in WBS definition, cited in the previous chapter, and the general rule of managing at a level at which you can reasonably estimate and control the work. It also results in poorly defined time estimates and will likely result in over or underestimates.

When using a project scheduling software, the way to find orphaned or incompletely linked tasks on a network diagram is to identify them by looking for tasks that have no arrow-headed lines connecting them to overall diagram. They usually stand out as stranded tasks in the middle of the chart.

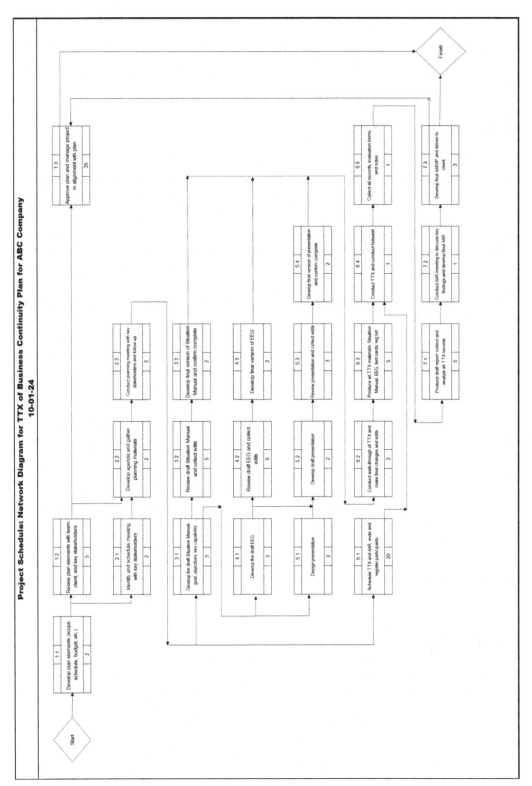

Figure 6.3 Project schedule-network diagram or TTX.

Once you develop some experience with developing and managing similar projects you can take a few shortcuts, and one is to still go through the process of developing your network diagram, but to manage a milestone schedule, especially when displaying project schedules for senior executives who want the general schedule vs. a detail of the whole schedule.

DEVELOPING TASK DURATIONS

Once you have developed a solid network diagram of project tasks, the next step is adding the estimated time durations for accomplishing those tasks. A time estimate is developed for each task duration. There are two essential ways to look at what creates time estimates for work tasks: fixed duration tasks or level of effort-driven tasks. When a task has a fixed duration, no matter how many resources that we add to it, it will always take a certain time. For example, if we are building a set of concrete berms to protect from flooding, the concrete takes a set amount of time to cure, and no matter how many resources we have, we cannot change the duration of that task (note – this is what we might call a buffer task, as this activity does not accumulate any costs direct; it can also be accounted for as a lag in project scheduling).

When task time estimates are variable by the level of effort applied to them, we consider them "effort driven". A clear example is "producing the draft report of the TTX". If we apply one person to complete this, then it will take a specific time. Add two people and we can possibly shrink the time to get this done, although it is unlikely to be a linear level of return on effort.

This is an important point to make at this time. There are some tasks which cannot be split between people that result in reduced work effort and significantly decrease the time to complete the task. Developing the plan documents would certainly benefit from having more than one person work on them, but there is a "law of diminishing returns" from adding more resources. Add two people to the task, and it may take less time, but add four people and the estimated work time is unlikely to decrease significantly. In fact, there may be detrimental effects of having too many resources on one task, the so-called "too many cooks in the kitchen" result, which may cause more work to coordinate efforts or may result in a less optimal outcome. This is true of a number of projects that require plan, training, or exercise development, although there are evident benefits from teamwork on specific deliverables.

TASK DURATION ESTIMATING

We have already done this in the Task Sequencing Chart, but the question arises, how do we develop time estimates for work? There are three basic methods for estimating task durations:

- **Rough order of magnitude (ROM):** Rough order of magnitude is usually used for estimating at the whole project level, where estimating is based on similar projects. However, this can also be used at the level of deliverables or objectives. An example of this would be taking a project that was used to implement a new Incident Command System (ICS) software at a state agency that, let us say, took 9 months to implement for 25 staff, and we then estimate that it will take 6 months for a 12-person county level emergency management agency. This may be based on a rough approximation of the equivalent time, not exactly a one to one ratio for a smaller project.

- **Parametric estimating:** Parametric estimating uses a unit of measure that is common to the industry or area of practice. This is what we call a "rule of thumb", and there are a number of parametric measures, such hours per lines of code in programming, hours for writing plans, and hours for specific types of construction work.[2]

  For example referring to Table 6.3, instructional developers might use a metric for eLearning development as given in Table 6.3 below:

  Ranges are provided as these are approximate measures. Developers will often use their experience and log their hours on projects to improve their estimating accuracy over time.

- **Bottom up (Detailed estimating):**

  Bottom-up estimating is a method in which a subject matter expert uses their knowledge and experience of the work to develop estimates using basic elements, identifying all resource needs: staff, equipment, supplies, systems (software), space, and services; and then estimating the hours it would take to complete the work on a task. This may include using detailed estimates from past project tasks or hours logged.

Table 6.3 eLearning Levels of Engagement

Level	Type of Interactions	Examples	Development Hours/1 Instructional Contact Hr.
1-Basic	"Page-turners" where learners click on the "next" button and navigation is limited. Includes basic images, simple animations and may include audio, and simple quizzes.	Most FEMA Independent Study Courses. See: https://training. fema.gov/is/	100–200
2-Intermediate	Contains more multimedia components, with expanded menus, glossaries and external links. May have limited animations and high-level assessments.	A Practical Guide to the Access and Functional Needs of Vulnerable Populations (offered by CDC through Tulane University)	200–300
3-Complex	Animated videos, audio, multi-screen scenarios that requires higher level cognitive skills to solve real-world problems and develop solutions.	Ready.gov Kids Disaster Preparedness Games www. ready.gov/kids/games	300–400
4-Games/ Simulations	Game-based interactions, exercises that simulate real-world environments with sophisticated evaluations.	America's Army, Enhanced Dynamic Geo-Social Environment (EDGE), Virtual Community Reception Center (vCRC; CDC)	500–700

- Here is an example using a task from above: *7.1 Produce draft report: collect and analyze all TTX records, evaluations, and other notes.* My analysis of the work required might be as follows:
  - Collect and analyze all TTX records – based on previous projects – 14 h.
  - Write first draft of AAR report – 8 h.
  - Review draft AAR report – 12 h.
  - Write second draft AAR report and review and proof – 6 h.
  - *Total = 40 h.*
- Not all of this may be direct working time as the person writing the report may have a meeting related to the work, may have to go back and check on some items, find other references, etc. The detailed estimates may be approximated here. As we mentioned before, the idea is to ensure that the total estimate is as close as possible. The project manager will manage the overall task, not the detailed work.

Before we move on to the next step, we should note that estimating methods for time are closely associated with those of costs. Naturally, a time a resource is used, regardless of the type, will impact the cost. As the saying goes, time is money. We will cover this in greater detail when we get to budgeting later in this chapter.

DEVELOPING THE CRITICAL PATH

Once we have come up with time estimates, we want to estimate the duration of the project using the *Critical Path Method*. Critical Path Method uses the network diagram, task relationships, and estimated task durations to compute the longest path through the network and hence the total project duration. It will also identify tasks on which there is slack (what PMI calls "float"), in other words time that the task may be delayed without delaying the entire project. Critical Path Method helps to identify those tasks which are most "critical" that can lead to delays on the entire project. It also helps to see where a project manager might add resources, find alternative methods and options, such as outsourcing, to reduce the overall project duration.

The steps in this process are to conduct a *forward pass* through the network diagram, and the second step is to conduct a *backward pass* through the network using the total duration. The forward pass will identify the total project duration, and the backward pass will identify the longest path or critical path through the network and all tasks which are not critical and have some slack time, the time it may be delayed without delating the entire project.

Figure 6.4 is a template for estimating four different items: Early Start, Early Finish, Late Start, and Late Finish, and the Slack. These are based on days, not on dates. That will come when we start converting the network diagram into a Gantt chart.

The Early Start and the Early Finish are the earliest a task can start and finish given the tasks that precede it. The Late Start and Late Finish are the latest a task can start and finish without delaying the overall project. The steps to construct a critical path are explained below.

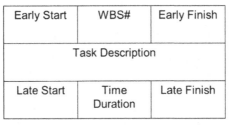

Early Start	WBS#	Early Finish
	Task Description	
Late Start	Time Duration	Late Finish

Slack (Float)

Figure 6.4 Critical path method task calculation.

STEP 1: FORWARD PASS

See Figure 6.5 using the TTX project.

1. Begin with the Start at 0 (zero). We start with zero because if we start with day 1 then in essence we are adding a day to the project. Since the Start is the only item before it, we carry over the 0 from the Start to the Early Start for task 1.1.
2. We add the two-days duration to the Early Start to get two for the Early Finish.
3. Since task 1.1 is the only one that precedes tasks 1.2 and 2.1, we carry the two days over from its Early Finish to each of their Early Starts. A reminder that we are working with durations here, not dates, so we do not add any days to the next Early Start times; otherwise we wind up adding a day each time we do this. You can think of this as a finishing point from the previous task.
4. Then we compute the Early Finish times for each of these tasks, 1.2 and 2.1, as you can see below by adding each of their task durations respectively to the Early Start. For 1.2 that would be ES 2+ 5 days duration=7. For 2.1 that would be ES 2+ 2 days duration=4.
5. Both tasks 1.2 and 2.1 precede task 2.2, so when multiple tasks precede another task, the rule we apply here is *Late Going Forward* when calculating the next Early Start. Given that the Earliest Task 1.2 finishes at the end of day 7, and the Earliest Task 2.1 finishes on day 4, we take day 7 as the Early Start for Task 2.2.

We do that for the rest of the network, calculating each of the task Early Starts and Early Finishes. If you skip ahead below, you will find the longest path and produce a total project duration is 42 days.

STEP 2: BACKWARD PASS

As we go through the steps for the backward pass, we will reference the section of the critical path chart represented in Figure 6.6.

1. We begin with the project finish and the latest Early Finish time among the tasks that precede it. There are two tasks which precede the finish, Tasks 1.3 and 7.3 which both have an Early Finish time of 42 days. We use 42 days as our Late Finish for each of the

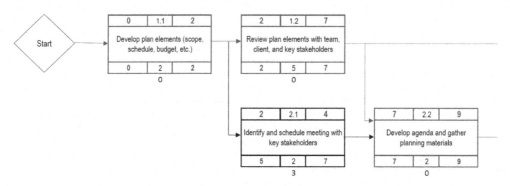

Figure 6.5 Critical path method – Forward pass diagram example.

preceding tasks and then begin working backward by subtracting the task duration to give us the Late Start time for each of them.

2. When a sequence of tasks has only one predecessor for each as you will see with Tasks 7.1–7.3, the Late Finish and Late Start times are the same as the Ealy Finish and Early times.

3. When we encounter a single task that has more than one successor, the rule we want to apply is *Early Going Back*. This is the opposite of the rule for the forward pass, *Late Going Forward*. In the table below you will see that Task 3.1 has two successor tasks, 3.2 and 4.1. Since 4.1 has a Late Start of 14 days which is earlier than the Late Start of 17 days, we take the earlier time. It is important to trace the paths backward appropriately to ensure that the logic is followed and the calculations are correct.

4. We continue working backward through the network diagram performing the backward pass, making sure to follow the paths, to calculate the Late Start and Late Finish Times for all tasks in the network diagram.

5. The next and final step is to calculate the slack on those tasks with a Late Finish time which is higher than the Early Finish. Slack is calculated by subtracting the Early Finish from the Late Finish. For example, Task 5.1 has a Late Finish of 17 and an Early Finish of 16, so there is one day of slack on that task. For ease of reference in our diagram, the number of days of slack for each task is given just under the box.

6. When you are finished, you should start to see a pattern, which is highlighted on the CPM diagram below: the path with lighter print is the critical path, while tasks in darker print have slack and are non-critical.

Note – Figure 6.7 and the figures given above are formatted to include all of the tasks and their relationships so they fit on the page. Typically, a network diagram is laid out from left to right from the Start, providing ample space to illustrate the sequential tasks relationships.

Refer to Figure 6.7. A couple of reminders when performing a forward and backward pass.

• Make sure that no tasks are orphaned. Any orphaned tasks may result in a miscalculation of the overall project duration and critical path.

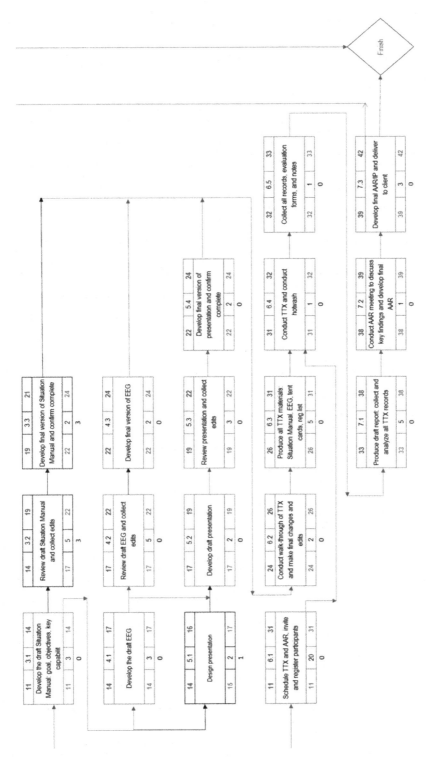

Figure 6.6 Critical path method – Forward pass diagram example – Project duration.

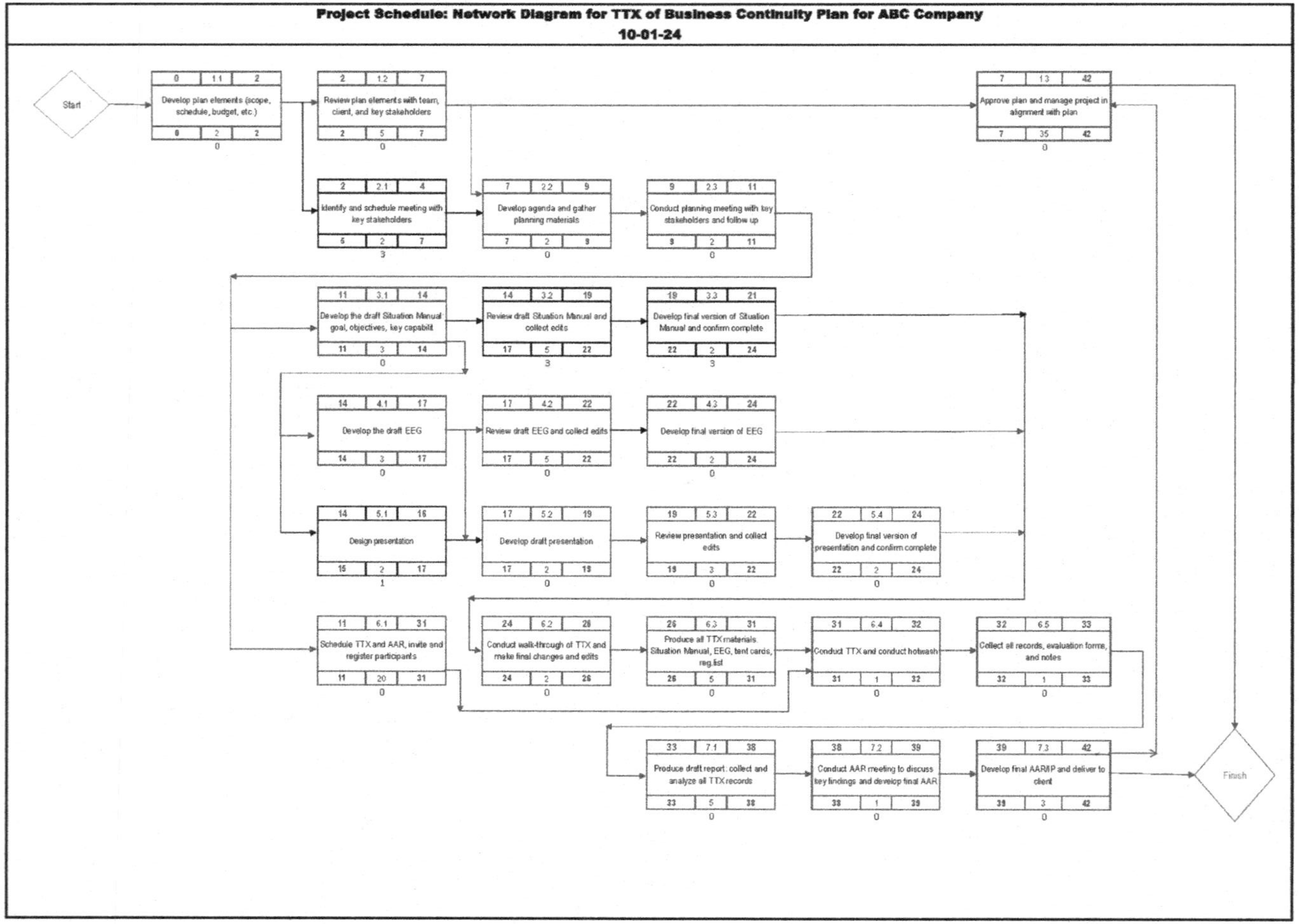

Figure 6.7 Critical path diagram for TTX.

- For any leads for a start-to-start relationship between tasks or lags with a finish-to-start time between tasks, it should be factored into the Early Start time.
- Remember that times are durations and not dates and that you need to use the calculations as they are and not make the mistake of starting the Early Start of a succeeding task on the next day, as this will add additional time into the schedule. Dates are accounted for in the step below.
- It is recommended that the schedule be drawn out as long and as clearly to reveal the logical path (examples above and below are only for the book).
- If you have network diagrams with a larger number of tasks (for example more than 50) and multiple paths, then you may want to include milestone reviews and to "sink" tasks (perhaps sync. is a better term; sink is the technical term) into those milestones to make it easier to analyze and calculate the schedule.
- To make it easier to performing calculations, round up to the nearest number instead of using fractions.

ESTIMATING TIME AND COST: OH IT IS SUCH AN ART

Before we go further I want to address some additional aspects of estimating. There is a clear relationship between time and cost estimating. Time is money as the saying goes, so the more time a resource (team member, equipment, etc.) spends on a task, the higher the cost. We covered a little bit of the science when we discussed the key three methods: ROM, parametric, and bottom-up estimating. However, estimates for time and cost are conditioned based on other factors, so there is an art associated with estimating, making estimating time and cost both a science and art.

To help solidify these factors, let me introduce you to the pneumonic for the following rules on estimating: *OH* it is such an *ART*:

Owner of task has responsibility
Human productivity
Accuracy depends on level of detail
Ranges with Risks
Trade off in terms of time/cost/ resource

Owner of Task Has Responsibility

As we mentioned before, if you are developing estimates, then the team members who are going to perform the work and are familiar enough with it (often called subject matter experts or SMEs) should calculate how long it will take. In essence, a team member should take ownership for the tasks they are performing or take part in. If a project manager develops the time and cost estimates without an SME for tasks they are unfamiliar with, then the team member can disavow the estimate. This is often a point of contention on projects, when I hear the question raised "who and where" estimates originated from, and the usual answer is "I don't know", or someone else (in another department or someone at a higher level, unfamiliar with the work). I make the point that this up to a team member, but it may also be a number of team members who might be involved in developing a task estimate. So, in short, this rule should make it clear that one person needs to be responsible for the task estimate, whether they develop it independently or as part of a team.

Human Productivity

People are only productive for a certain amount of time. They also need time to ramp up and to close things out when they work on a task, especially knowledge-based work; time to set their work up before they start full steam, and then, when they begin to close things out, check their work or put a placeholder before they close things out for the day or for an entire task.

Let's look at an example. If we look at a standard working day (in the USA) at 8 h for someone working in an office, that person may be productive for 5–6 h, not including time for lunch; there will be time for short breaks (including bio-breaks), meetings, minor interruptions, etc. And I know that there will be individuals who will read this and make the statement that they work 10–12 h days and the like, which is plausible, but not sustainable for many days on end. It results in burnout, even for the most robust performer at work.

Another aspect of human productivity is the Learning Curve Effect. The Learning Effect is the phenomenon that as we perform work we get better at it, identifying improvements that create greater efficiencies as time goes on. The chart below is an illustration of the learning curve, with the learning (productivity) and experience, over time on the Y axis, for two products; Product A and Product B. This can be applied to any number of different types of work (Figure 6.8).

There are also other learning curve effects associated with working with any novel content, organizational, or environmental factors which may be new to a project team or team members, especially if they are a hired contractor working with a new client or working with new team members. We will explore this further in the next chapter when we cover a managing team and team formation.

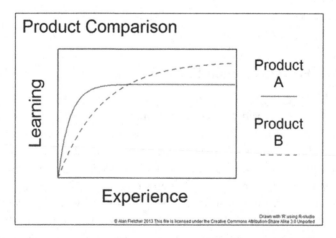

Figure 6.8 Learning curve effect. (Credit: Alan Fletcher, CC BY-SA 3.0 DEED Attribution-ShareAlike 3.0 Unported.)

Accuracy Depends on Level of Detail

Time and cost estimates are more accurate the more relevant details are collected on the task. For example, if we only know that in our project example, developing a TTX Business Continuity ABC company, that for Task 2.1 we are identifying and organizing the stakeholder meeting without any idea of how many stakeholders there are, where they may be (local, or dispersed: national, international), and how available they are, then our estimate would reflect this ambiguity. On the other hand, if we know that there are 15 major stakeholders, who are all working in the same location, and that they are available in the approximate range of time planned for our key meetings and TTX event, then we can provide a more precise estimate.

Ranges with Risks

Time and cost estimates have risks associated with them, both positive (opportunities) and negative. We usually associate negative aspects with risks, so let's start there. For any team member who submits a time estimate, there is a risk that they may not be able to perform it within the time frame given (for a variety of reasons: sickness, other assignments are given priority, weather-related risks, etc.). There are personal sanctions they might encounter if they fail to deliver on time: loss of reputation, denied career opportunities, or, in very extreme situations when the project stakes are high, a demotion or dismissal.

Individuals on a project team may also be overly optimistic or pessimistic when it comes to estimates. They may be overly optimistic (providing lower time estimates) to please the "boss" or higher ups, if that is what they are looking for (or demanding); or pessimistic if they wish to protect themselves, if they fear a downside as presented above. The same is true for the overall project schedule; often the key stakeholder expectations, client or project sponsor, might drive the project manager to develop an entire schedule that is overly optimistic based on artificial deadlines, sometime according to bid or grant funding deadlines.

One final thought on this, and that is Parkinson's Law: work takes up the volume of time that it is given. As discussed above, much of how a project team member might estimate the time it would take to complete is based on the organizational culture, and whether they feel comfortable providing a more ambitious estimate, which might challenge them to get it done quicker; or that they would feel a potential downside for themselves and provide a more conservative estimate. At the end of the day, as much as a project manager or the project champion might push team members, we still need to take into consideration what is a realistic timeframe for getting the work done.

Trade-Off in Terms of Time/Cost/Resource

When it comes to resource allocation, and the time and costs associated with accomplishing a task, there are different trade-offs to consider:

- Deciding on what types of resources, adding higher qualified more expensive vs. less qualified less expensive staff resources to a task.
- Adding more staff members to a task, with the potential for diminishing returns.
- Adding specialized equipment or systems/software to improve productivity.

For example, there are some tasks which may benefit from having more than one person work on them. *Task 7.3 Develop final AAR/IP and deliver to client* might be done by one person and take five days, or we might apply two people to the task, one to work on the writing and editing, and another to manage the formatting and any graphics to reduce the time to three days.

One question that is often raised is whether to account for internal staff time. Often, the assumption is that the time staff members, internal to the organization, are working on the project is already paid for, so it does not worth making an effort to account for their time. However, this is misconception of the true time and cost for a project. The time staff are working on a project is an opportunity cost, as time spent on one project is time they may be using to work on other assignments or other projects. If that other work, either on other assignments (operational work) or other projects, it means it does not get done or is delayed, and there is a cost for that.

While it may take more work, it is a better practice to include all time for both external and internal staff, including any time taken for meetings with stakeholders, into consideration for the schedule and budget. In addition, this also benefits the program and organization as a whole as the true cost of a project can be used to measure one project against another, as well as evaluating whether it may make sense to outsource a project or to develop it internally.

NEXT STEP: DEVELOPING A GANTT CHART AND CALENDAR

Now we are on to the next step in developing the schedule which is to take the network diagram and add resources (people, equipment/systems, facilities, and materials/supplies), and specific dates to the schedule. We need to take into consideration resource availability and identify non-working time, specific to individual project team members or the organization. We do this by looking at the calendar and identifying:

- Major holidays
- Major milestone dates, including reviews, inspections, audits, or other potential interruptions
- Vacation days for team members
- Estimated number of sick days based on historical data (or other factors)
- Any other non-working times

We then block out all non-working time on the calendar. It is very important to take this into consideration as you map out a project schedule, taking into account, not just vacation days, sick days, important work events, but also the unforeseen incidents that may impact the project timeline. When you work in a high-tempo organization that responds to emergencies small and large, schedules often get dashed, so once we have performed a project risk analysis (covered in the next chapter), we might allocate time buffers as appropriate.

The final step is adding these to a Gantt chart by sequencing each of them by date, using the Critical Path diagram. In Figure 6.9 is an example of Gantt chart using a standard project management software.

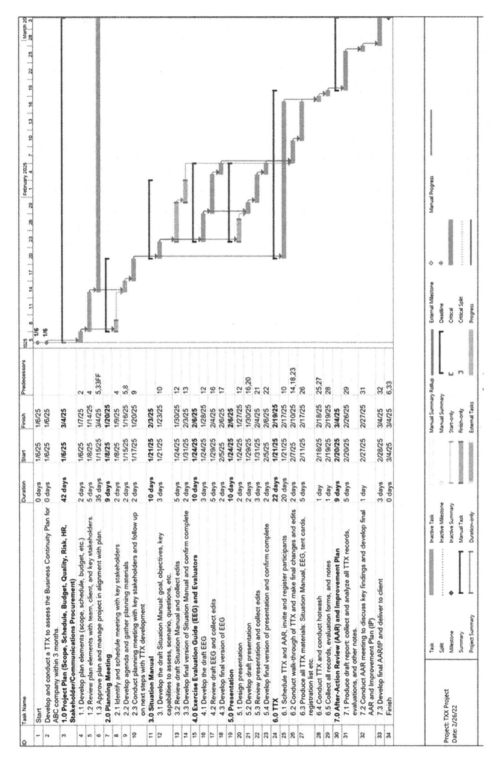

Figure 6.9 Gantt chart for TTX-BCP for ABC.

PROJECT MANAGEMENT SOFTWARE

Over the years that I have been teaching project management and guiding project managers, the question often arises: project management software performs these calculations for me, so why do I need to know how to construct a network diagram and use critical path method (CPM)? The short answer, it is important to understand the mechanics of CPM calculations; if you don't understand them, then you won't know if the calculations are wrong or why they may be wrong.

This is much like learning the basics of accounting is necessary before you start using spreadsheets to facilitate accounting transactions on a profit and loss statement or a balance sheet. Project management software can help with the calculations of the critical path, but learning to use the software does not teach foundational project skills. Depending on how complex the project and schedule are, I encourage beginners to use a spreadsheet program to create a schedule; for more complex schedules and skilled project managers, I recommend exploring different project management software programs that suit the needs of the project.

NOTES

1 The Gantt chart developed by Henry L. Gantt in the early 1900s as part of his work for various industrial companies, among them Bethlehem Steel. He also helped to improve production scheduling as part of the war effort in the USA for World War I. See: www.bl.uk/people/henry-laurence-gantt
2 Many years ago, I developed two separate classroom-based courses in Construction Cost Estimating and Construction Scheduling, respectively. The RS Means series of books provides indices for labor hours for a wide range of construction activities, used by many PMs in the construction field. See: www.rsmeans.com/products/books

Chapter 7

Developing the Project Budget

Managing a budget is like a tightrope act. They require careful balance and a touch of creativity to stay on track.

Peter Drucker, Management Consultant and Author

COST MANAGEMENT: BUDGETING, CASH FLOWS, AND GOVERNMENT FUNDING CYCLES

We develop cost estimates based on the work defined in the WBS, including the time estimates from the schedule development process. We then assign resources to tasks:

- Human resources: team members assigned on a full-time basis (maybe over weeks) or part-time basis; contract or seconded staff
- Equipment: heavy equipment (trucks, forklifts), computers, servers, networks, radios and base stations, etc.
- Systems and software
- Materials and supplies
- Locations: venues used for meetings, exercises, or rented office space for work; or warehouse storage, staging areas, or operations centers
- Services, for example, vendors who might set up facilities or third-party logistics (3PL) providers separate from materials

We can then analyze the work duration or level of effort needed for tasks to calculate labor hours, unit costs for equipment or services (if based on an amortization basis), and materials, supplies, and services that are consumed in those tasks.

Let's take a look at one task to see how this works in practice. We will use task 2.3 *Conduct planning meeting with key stakeholders and follow up on next steps with TTX development.*

The first step is to identify the resource costs by staff and non-staff; see an example in Table 7.1.

You will notice that costs are identified by hourly rate, which include overhead and fringe costs. These are also summarized under teams associated with three teams: Internal Exercise Team, Exercise Development Team, and Evaluation Team. Non-staff resource

DOI: 10.1201/9781003201557-9

Table 7.1 Task Budget Estimates

	Position		Hr. Rate Staff	Hr. Rate Total Per Team
Internal Exercise Team				$ 775
Carol Lee	Director of Emergency Management		$ 150	
Robert Carnevali	Director of Business Continuity		$ 125	
Tanisha Walker	Director of Training and Exercises		$ 125	
Sandra Hernandez	Manager IT Security		$ 175	
Victoria Lee	VP of Human Resources		$ 200	
Exercise Development Team				$ 350
Chad Kim	Developer/Emergency Manager		$ 125	
Susan Shields	Project Manager		$ 150	
Alex Margolis	Editor		$ 75	
Evaluation Team				$ 450
Maria Kelly	Emergency Management Professional		$ 150	
Jack Torres	(former) VP of Business Continuity		$ 200	
Hae Sun Kim	Continuity Professional		$ 100	

Equipment Unit Costs	Cost Calculation	**Total Cost**	**Unit Cost/ Day/Task**	
Laptop computer	Amortization	$ 1,500	$ 5.75	
Printing	Based on 70 leaves/35 pages × # TTX participants	NA	$ 45	
Training and exercise room	Daily rental cost	NA	$ 800	

costs have been identified in a per unit cost. This will make it easier for us to apply to tasks by allocating them based on the total number of hours by staff, team, and/or unit costs for non-staff resources.

Now let's walk through each of these calculations, starting from the right and going left given in Table 7.2.

- The Training and Exercise Room costs $800 per day to rent and can host all 11 staff who will be attending.

Table 7.2 Task Budget Estimate Calculation

Task# and Description	Task Duration Days	Total	Staff Assigned			Non-Staff Resources		
			Ex. Dev. Team	Internal Exercise Team	Eval. Team	Laptop Computer	Printing	Training and Exercise Room
2.3 Conduct planning meeting with key stakeholders and follow up on next steps with TTX development	2	$ 16,482	$ 5,600	$ 6,200	$ 3,600	$ 34.48	$ 247.50	$800

- Printed materials such as copies of the agenda and project plan documents, Business Continuity Plan, and other documents are estimated at half the cost per unit $22.50 × 11 staff members.
- Laptop costs are based on the amortization cost per day $5.75 × 2 days × 3 staff members on the Exercise Dev. Team.
- Evaluation Team cost is 8 h × $450/h.
- Internal Exercise Team cost is 8 h × $775/h.
- Exercise Development Team cost is 2 days × 8 h × $350/h.

Task 2.3 Total cost= $16,482

Notice that the only cost computed based on the task duration is the Exercise Dev. Team, as we assume they will be working on developing and planning the meeting, so we account for those additional hours. All other costs are based on a full day for the TTX planning meeting.

We would walk through the rest of the tasks, allocating and calculating costs with assigned resources, and what we would end up would look like an example in Figure 7.1. Many organizations use accounting codes or set budget line items for specific costs, so for budgeting purposes you should obtain that list and budgeting format. Using a method adapted from the one presented above and then correlating it to the budget lines will help you trace how costs are calculated and track these along budget lines.

A couple of items to note when it comes to project costs and budgeting, in particular for grant funded projects are as follows:

- For multi-year projects you need to take into consideration inflation and cost of living adjustments/raises etc.
- The need to account for any changes in market prices for goods and services, rentals, etc.
- Holding costs for any rental equipment that you need to have on hand before you need it. This needs to take into consideration lead time for shipping, setup, any system checks, and breakdown and return to the vendor or owner, as well as insurance.
- Overhead costs and fringe benefits; there may be standard rates for what is billed to projects in your organization.
- Indirect costs for your organization or company; these are usually standardized by governmental grant programs, with limits set; there are also rates set by organizations

Tasks and Descriptions	Task Duration Days	Total	Ex. Dev. Team	Internal Exercise Team	Eval. Team	Susan Shields	Carol Lee	Chad Kim	Robert Carnevali	Alex Margolis	Laptop Computer	Printing	Training and Exercise Room
						Staff Assigned					Non-Staff Resources		
1.1 Develop plan elements (scope, schedule, budget, etc.)	2	$ 2,434				$ 2,400					$ 34.48		
1.2 Review plan elements with team, client, and key stakeholders.	5	$ 2,899	$ 875	1,938							$ 86.21		
1.3 Approve plan and manage project in alignment with plan.	35	$ 15,478				$ 5,250	5,250		$ 4,375		$ 603.45		
2.1 Identify and schedule planning meetings with key stakeholders	2	$ 2,434				$ 2,400					$ 34.48		
2.2 Develop agenda and gather planning materials	2	$ 2,834	2,800								$ 34.48		
2.3 Conduct planning meeting with key stakeholders and follow up on next steps with TTX development	2	$ 26,034	5,600	12,400	$ 7,200						$ 34.48		$ 800
3.1 Develop the draft Situation Manual: goal, objectives, key capabilities to assess, scenario, questions, etc.	3	$ 8,452	8,400								$ 51.72		
3.2 Review draft Situation Manual and collect edits	5	$ 6,286		6,200							$ 86.21		
3.3 Develop final version of Situation Manual and confirm complete	2	$ 1,434	1,400								$ 34.48		
4.1 Develop the draft EEG	3	$ 2,452				$ 1,200		$ 1,200			$ 51.72		
4.2 Review draft EEG and collect edits	3	$ 5,486			$ 3,600	$ 1,200				$ 600	$ 86.21		
4.3 Develop final version of EEG	2	$ 1,234				$ 600		$ 600			$ 34.48		
5.1 Design presentation	2	$ 634				$ 300		$ 300			$ 34.48		
5.2 Develop draft presentation	2	$ 934				$ 450		$ 450			$ 34.48		
5.3 Review presentation and collect edits	3	$ 752	700								$ 51.72		
5.4 Develop final version of presentation and confirm complete	2	$ 1,234				$ 600		$ 600			$ 34.48		
6.1 Schedule TTX and AAR, invite and register participants	20	$ 6,345				$ 3,000		$ 3,000			$ 344.83		
6.2 Conduct walk-through of TTX and make final changes and edits	2	$ 4,234	4,200								$ 34.48		
6.3 Produce all TTX materials: Situation Manual, EEG, test cards, registration list etc.	5	$ 2,331				$ 1,200					$ 86.21	1,045	
6.4 Conduct TTX and conduct hotwash	1	$ 13,417	2,800	6,200	3,600						$ 17.24		800
6.5 Collect all records, evaluation forms, and notes	1	$ 1,417	1,400								$ 17.24		
7.1 Produce draft report: collect and analyze all TTX records, evaluations, and other notes.	5	$ 3,111	2,800								$ 86.21	225	
7.2 Conduct AAR meeting to discuss key findings and develop final AAR and Improvement Plan (IP)	1	$ 7,117	1,400	3,100	1,800						$ 17.24		$ 800
7.3 Develop final AAR IP and deliver to client	3	$ 1,992	1,400								$ 51.72	540	
Total		$ 120,980											

Figure 7.1 Project budget for TTX of business continuity plan for ABC company.

(universities, corporate) for what they usually charge for managing grant funded projects. Quite often these may be negotiated.

- It is a better practice to keep a small reserve of about 5% of the total budget for any unforeseen costs or overages.
- For government grant funded projects that stipulate the constraints of what may be spent within line items, ensure you have some degree of line-item flexibility to spend funds between those line items, for example, between personnel expenses and "other than personnel expenses", i.e. equipment, vendor services, materials, and travel.
- Budget for the true admin. costs, independent of other projects. When it comes to multiple projects with the same client or funding stream, if a project assumes that some of the admin. costs are covered by another project, then write that into the project and make that part of a periodic review.
- Where it is possible (with emphasis, as this is not always allowed by funding organizations or by the implementing organization, both for-profit, non-profit, and educational organizations), increase indirect cost rates to cover:
  - start-up costs: organizations (most often non-profit/educational entities) often begin by funding projects from existing reserves and then paying them back once they collect funds.
  - marketing costs: if your organization is doing great work that your clients are happy with, then why not fund that work (you will some other way).
  - closing costs: handover, maintenance, and support – while this should be a direct cost incorporated into the second bullet item, multi-year budgets that are fully funded are a rarity.

CASH FLOWS, GOVERNMENT FUNDING CYCLES, AND FINANCIAL ACCOUNTING SYSTEMS

Managing cash flows, when money flows from one bank account to another making it available to spend on project-related costs and accounting for those funds is critical to a project's success. Making sure you have the money available and can buy or pay for staff, vendor services, equipment, etc. can ensure timely starts and completion times. Many projects (some might say most) can suffer from funding delays or delays in funding allocation by fiscal authorities, which may result in late project starting times, funding availability, and, in some circumstances, funds being withdrawn when they have not been spent. It can also impact project team morale, vendor morale, and availability of resources (the old adage "you snooze, you lose").

While some grants only require a budget covering the entire specified time period, a year or shorter, some grant funded projects (request for proposals, bids, etc.) require a budget statement by month, quarter, or over a number of years, so that the donor and/or the fiscal agent (finance office/department) can encumber and allocate funding into accounts periodically, essentially managing what are the cash flows covering the project.

Budgets are also usually set according to a fiscal calendar according to the donor agency. The fiscal calendar or year may not be in alignment with a calendar year. For example, in New York City the fiscal year begins on July 1 and ends on June 30th of the following year. What this means is that any contracts/subcontracts are funded accordingly with period of

Table 7.3 Monthly Cash Flow Projection

Task# and Description	Total	January	February	March
1.1 Develop plan elements (scope, schedule, budget, etc.)	$ 2,434	$ 2,434		
1.2 Review plan elements with team, client, and key stakeholders	$ 2,899	$ 2,899		
1.3 Approve plan and manage project in alignment with plan	$ 15,478	$ 15,478		
2.1 Identify and schedule planning meetings with key stakeholders	$ 2,434	$ 2,434		
2.2 Develop agenda and gather planning materials	$ 2,834	$ 2,834		
2.3 Conduct planning meeting with key stakeholders and follow up on next steps with TTX development	$ 16,482	$ 5,494	$ 5,494	$ 5,494
3.1 Develop the draft Situation Manual: goal, objectives, key capabilities to assess, scenario, questions, etc.	$ 8,452		$ 8,452	
3.2 Review draft Situation Manual and collect edits	$ 6,286		$ 6,286	
3.3 Develop final version of Situation Manual and confirm complete	$ 1,434			$ 1,434
4.1 Develop the draft EEG	$ 2,452		$ 2,452	
4.2 Review draft EEG and collect edits	$ 5,486		$ 5,486	
4.3 Develop final version of EEG	$ 1,234			$ 1,234
5.1 Design presentation	$ 634		$ 634	
5.2 Develop draft presentation	$ 934		$ 934	
5.3 Review presentation and collect edits	$ 752		$ 752	
5.4 Develop final version of presentation and confirm complete	$ 1,234			$ 1,234
6.1 Schedule TTX and AAR, invite and register participants	$ 6,345		$ 6,345	
6.2 Conduct walk-through of TTX and make final changes and edits	$ 4,234		$ 4,234	
6.3 Produce all TTX materials: Situation Manual, EEG, tent cards, registration list, etc.	$ 2,331		$ 2,331	
6.4 Conduct TTX and conduct hotwash	$ 13,417			$ 13,417
6.5 Collect all records, evaluation forms, and notes	$ 1,417			$ 1,417
7.1 Produce draft report: collect and analyze all TTX records, evaluations, and other notes	$ 3,111			$ 3,111
7.2 Conduct AAR meeting to discuss key findings and develop final AAR and Improvement Plan (IP)	$ 7,117			$ 7,117
7.3 Develop final AAR/IP and deliver to client	$ 1,992			$ 1,992
	$ 111,428	**$ 31,575**	**$ 43,401**	**$ 36,452**

performance and payments that align with those dates. Managing these contract/subcontract timelines is critical to ensuring that funding is spent down and that deadlines are not exceeded which fall outside of these funding deadlines.

As presented above, you can use a the schedule, correlating the budget according to the expected timeline. See an example in Table 7.3 of the TTX Project with a budget allocation by month.

All organizations use an electronic accounting system to manage grant funded projects, and organizational accountants and financial managers are the professionals who manage these systems. Knowing key aspects of these systems, such as the accounting or budget lines these systems use and how they may influence your project is important. It is also critical to identify what forms or approvals are needed at departmental levels for your budget and spending process, and how these are used for reporting purposes.

Based on my years of experience in managing project budgets and with conversations with financial professionals and other project professionals, especially with government grants, here are a couple of final words on budgeting and financial management:

- Get to know your finance management professionals (financial controller, fiscal agent; whatever title they might hold, and it may be more than one person; bottom line is: the money manager) and explain the purpose of the project, present them with the estimated cash flows, ask what type of reports are available for your project and how long cash allocations and contracting may take (a conversation that may require legal counsel as well).
- Do not proceed without having money in your organization's bank account or formal approval in writing to spend from your project.
- If approval has not been given and costs are imminent, give warning that you will either not start or stop work at least one month prior to incurring the expenses.
- Make sure all contract dates and budget spending adheres to the fiscal year.
- Plan for a regular review with your financial manager, monthly or as necessary, to see where the project is according to the budget.

We will delve further into managing and controlling the project budget in Chapter 13.

Chapter 8

Developing the Human Resources Plan

> Whatever we do must be in accord with human nature. We cannot drive people; we must direct their development. The general policy of the past has been to drive; but the era of force must give way to that of knowledge, and the policy of the future will be to teach and lead, to the advantage of all concerned.
>
> *Henry L. Gantt, M.E.*

We have covered a great deal of the mechanics of the deterministic approach to project management (agile and critical chain being other variations that we will address later in the book). Now we turn to what I call the other project plan components or elements: HR, quality, risk, and communications. We will address procurement in a later chapter as well.

People unfamiliar with the professional practice of project management often perceive a project plan as simply the scope, schedule, and budget. This misses the mark of what makes project management a true art and into an integral whole. A well-constructed project plan needs to incorporate many elements for it to be successful:

- Team dynamics and stakeholder influences.
- Quality requirements and processes, and the related aspects of....
- Risk management.
- Managing communications with the team and stakeholders to understand and manage their expectations, and what the team can deliver, in order to ensure they are all on the same page and can adjust to any needed course corrections.

As the quote above references, taken from the time of mechanistic approaches to management,[1] we need to plan for these other elements by working together with the team, with client teams, end-users, and vendors/contractors, and other stakeholders to connect these pieces and to integrate them with one another for a comprehensive, whole, realistic, and effective plan. In this chapter we will refer to a sample project designed to create a web-based application to manage a group of emergency shelters to house residents evacuated from their homes.

HR MANAGEMENT: ASSEMBLING THE TEAM

The first phase in human resource management is assembling and developing the core project team. On larger projects there may be several teams that form the larger team, such

DOI: 10.1201/9781003201557-10

as the client team, a vendor/contractor team (may be more than one), and the core development team. The core development team comprises key functional areas of the project, which will vary according to the type of project. Development teams may comprise several teams on much larger development projects, especially in the fields of engineering where each function, such as excavation, construction, logistics, and safety, to name a few, may constitute separate teams.

By identifying the work required for the project and key functional knowledge areas, you will identify the expertise and skills required for the project team. For example, on a new systems development project, in addition to a project manager, the teams may comprise a lead developer, several junior developers working under the lead developer (application architect, database developer, etc.), Q/A and Testing Manager, UX Designer, and may include a separate Training and User Documentation Developer. One very common way to identify and illustrate who is needed on a project team is an organization chart. See an example of a software application project organization chart in Figure 8.1.

The organization chart illustrates the organizational connections by solid lines within the IT department where the software development team reports into the Project Management Director who reports up to the IT Chief Officer (most commonly referred to as a CTO), and also where related support units report up within their reporting lines. This chart also shows the connections to other parts of the organization, Finance, Human Resources, and Procurement, as well the relationship to the client team, demonstrated by a dotted line.

Another method to identify team members and staffing resources needed that is commonly used is the responsibility assignment matrix (RAM). The RAM is designed to show what human resources are assigned to each major project deliverable (or a functional area) and/or detailed down to the task level and the role that they play. An example with a smaller section of the RAM following from the same software development project based on major deliverables is given in Table 8.1.

Note – due to limited page space, only some of the deliverables are included in this sample RACI chart.

This version of the RAM is commonly referred to as a RACI chart as it stands for the designated role on the project:

- **Responsible** – the person who is responsible for carrying out the task. I recommend that one person be assigned as responsible.
- **Accountable** – the person to whom it is accountable, who authorizes or approves whether the deliverables, objective, or task have been completed. Usually this is one person.
- **Consult** – a person who consults on the task, providing their expertise or helping to do some, but not all of the work.
- **Inform** – a person who is informed on the task progress but not involved directly in the work.

Aspects of the RAM, such as who is assigned to a task, are commonly incorporated into Gantt charts (and project software: MS Project, ProjectLibre, Trello, Miro, etc.), but they often do not reflect the project role completely, so a stand-alone RAM is good tool to reflect who is responsible for a task, who is accountable to whom, and who is involved. As you can see in the example above, only one team member is Responsible and one is

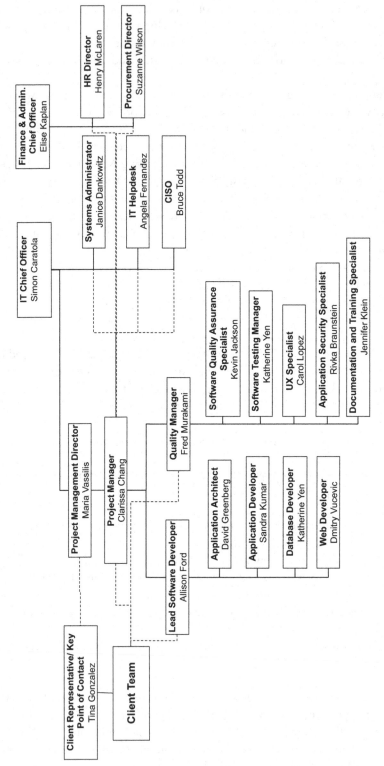

Sample- Web-based Software Application Project Team: Organization Chart

Note: this organization chart is for illustration purposes only and is not intended to reflect exactly how a software development team should be developed. All names used are fictional and are not intended to represent any past or current projects.

Figure 8.1 Project organizational chart for a software application team.

Table 8.1 RACI Chart for Web-Based Software Application

Project Deliverable	Maria Vassilis, PM Dir.	Clarissa Chang, Project Mgr.	David Greenberg, App. Architect	Sandra Kumar, App. Dev.	Katherine Yen, Database Dev.	Dmitry Vucevic, Web Developer	Kevin Jackson, Software Qual. Assurance	Katherine Yen, Software Test Mgr.	Carol Lopez, UX Spec.	Rivka Braunstein, App. Sec. Spec.	Janice Dankowitz, Sys. Admin.	Client Team
Software Requirements	I	A	R	C	C	C	C	I	C	C	I	C
Server Infrastructure	I	A	C	C	C	C	I	I	I	C	R	I
Web Application	I	A	C	C	C	R	I	I	C	C	I	I
Backend Database	I	A	C	C	R	C	I	I	I	C	C	I

Accountable for each major deliverable. You may also notice that some team members, including the client team, may consult or be informed. This can be useful for deciding what meetings and critical check-ins, including milestone reviews, that team members need to be involved in.

Between the WBS, the RAM, and an organization chart, and determining the estimated level of effort in work hours for each task, you should be able to determine whether project team members may be needed for the full duration of the project, that is full time; or if they may be needed only for temporary (short term), or part-time basis, what are called in management budget as F/T or P/T equivalents.

There may also be part-time staff sources from other parts of the organization, such as a financial accountant, IT systems administrator, or HR managers, who provide additional project support. Project team members are assembled in a variety of ways:

- Project team members may be pre-selected from within the organization by the project champion or senior management leadership.
- Project team members may be selected or recruited by the project manager from within the organization (more ideal than that above).
- Project team members may be newly hired from outside the organization (also ideal, but time-consuming as it requires working with the organization's HR department for recruitment, hiring, and onboarding).
- The project manager may have to bargain for the talent and their time needed to plan and carry out the project. Alas, this often happens on small projects and in organizations which undervalue project management as a discipline. This is less than ideal and often assumes that staff time is an assumed cost (or perhaps a sunk cost) and that it is just additional responsibilities that organizational staff may take on. However, this is not the case as there is a trade-off when staff are working on a project and not on other organizational work.

Concerning this last point, it is important to account for the time worked on a project, whether organizational staff are salaried employees or contract staff, as this accounts for the true costs accrued to the project. This includes the time devoted to the project by any senior management or support staff, such as finance and human resources staff. Identifying and managing attributable worked time and related costs ensures fiduciary responsibility for a project and allows for clear measures of project success and comparison with other projects.

ESTABLISHING AND DEVELOPING THE PROJECT TEAM

Now that you have identified a core project team, you want to help develop them into a high-performing team, especially if you are going to be working over a period of months or longer. Some team members may be familiar with one another, but, quite often, project teams are assembled from a variety of different organizations or with contract hires to develop plans, training, exercises, and systems. All projects with new team members typically go through the four phases of team development by Bruce Tuckman.[2]

Refer to Figure 8.2. These stages are characterized by:

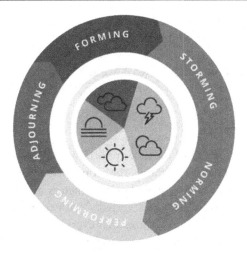

Figure 8.2 Phases of team development. (From DovileMi – Own work, CC BY-SA 4.0, https://commons.
    wikimedia.org/w/index.php?curid=101823819)

- **Forming** – In the initial stage team members meet and become acquainted with the scope of the project, their roles and responsibilities. Team members are usually excited and motivated about the project, but anxious about their new endeavor and how they fit in.
- **Storming** – In this stage team members test boundaries, challenge authority, and may engage in confrontation with other team members. There may be work imbalances or communication issues that crop up at this stage, leading to some underlying tensions and lower motivation.
- **Norming** – At this stage team members start to negotiate their conflicts and resolve their differences, better understanding their role, responsibilities, and collaboration with other team members. Motivation is higher among the team.
- **Performing** – At this stage group roles and the norms have been established, and team members are highly motivated and performing at their roles and working well together.

There are steps that project managers can take to ensure that new team members pass through these stages effectively to achieve higher levels of team performance.

HIGH-PERFORMANCE TEAMS

In their book *The Wisdom of Teams: Creating the High-Performance Organization,*[3] Jon Katzenbach and Douglas Smith identified the qualities that high-performance teams tend to exhibit

- Strong sense of purpose, with mutual accountability toward a common purpose.
- Ambitious performance goals compared to the average teams, in general.
- Effective plans and operational procedures, with a common understanding around plans and procedures.

- Individual obligations to their specific roles.
- Complementary skill sets, and interchangeable skills.

Let's unpack some of these in terms of what we have covered so far and what we will cover.

Strong Sense of Purpose

It has been my experience that one of the first important things when starting a project is to ensure that every team member understands the scope of work, in particular the goal and major deliverables or objectives and other key parameters and requirements. Reviewing this both on paper (or electronically) and verbally with team members, and having an open discussion and taking questions, ensures that everyone is on the same page. Every so often it may help to remind the team of the project goal and touch upon some aspect of it when addressing agenda items in a meeting or check-in with team members.

Ambitious Performance Goals

My general impression is that professional, capable project and emergency managers with whom I have worked over the years are ambitious in their careers, often what are called "Type A" personalities who exhibit a competitiveness and tenacity in achieving their goals. In the sphere of emergency management and public safety, where the main aim is to preserve and protect people's lives and communities, there is an inherently strong drive to get the job done and to perform at the highest level. Project managers can inspire high performance by highlighting this life-safety and security mission at the core of all projects addressed in this book and addressing what motivates the individual team members.

Effective Plans and Operational Procedures

As we cover in this book, there are better practices to follow when it comes to project management. The same is also true when it comes to emergency management as evident by ICS, NIMS, EMAP, and the NFPA standards, and other emergency standards. This is also true when it comes to other professional disciplines, whether it is software development, graphic production, course authoring, or engineering; each has its own professional practices, and effective managers of all types tend to choose team resources (internal, contractors, and vendors) who are reliable and solid. Ultimately, it is the project manager's responsibility to ensure that each team member is able to carry out their work according to their respective professional practices and know when and how they integrate their work with other team members. Project managers who follow better practices and ensure they are followed, knowing what quality work is expected, help to ensure effective team performance.

Individual Obligations to Specific Roles

The first thing a project manager needs to make sure of is that each team member knows their role and position expectations and how they integrate with the team. In some cases, you may need to provide a job description or a position's initial scope of work to make sure there is a mutual understanding of what is required, especially for new team members.

The RAM discussed earlier in human resources management can be effective tools in communicating roles and responsibilities. This can also help coordinate who is needed on what tasks and at what times to help with scheduling.

Successful project teams exhibit qualities such as being responsive to other team members on their work, showing up to meetings and on time, and following through. There is a strong sense of mutual ownership when you hear a high-performance team member speak about their project work. Project managers need to keep their team members accountable for their work, not micromanaging, but rather managing to plan objectives, outcomes, and requirements.

Complementary Skill Sets and Interchangeable Skills

While it is not present or possible on all projects, some project teams, especially in emergency management and public safety, have some degree of complementarity, shared competencies, and may be interchangeable. This is often the case when it comes to those who are familiar with ICS and have years of experience in the field, with management of planning, operations, logistics, and/or finance and administration. It may not always be that each team member is able to perform at an equal level as the person they may fill in for, but the team collectively often rises to the occasion, much like a sports team to fill the gap. Creating a team that has complementary and interchangeable skills starts with team selection and identifying the knowledge, skills, and abilities that overlap in critical areas required for the project scope.

TEAM ADMINISTRATION, MANAGEMENT, AND RULES OF ENGAGEMENT

While every team may not be a high-performing team at first, as discussed with new teams, getting there requires providing some clear rules of the road and the "vehicle" for the journey. Some of these recommended actions are identified above, and we will discuss project leadership at greater length in Chapter 14. When it comes to managing individuals or teams in the workplace, there are numerous great books on the fundamentals on management. I highly recommend reading some of those books and also taking courses on management when appropriate. I will highlight some of the basic elements that are needed as a minimum to maintain a well-functioning team.

CREATING AND MAINTAINING PLANNING/COORDINATING COMMITTEES

Many government-funded projects, and at times corporate and non-profit organizations, create a project steering committee, project review board, or similar group that will oversee the project on a periodic basis, perhaps once a month, quarterly or less often on longer term projects. These types of committees are empowered to review project status, hold the project manager and team accountable for progress, provide approval at project stage gate/milestone reviews, and to provide guidance on any changes to the corporate strategic objectives, industry and company compliance, and communication from the client's senior leadership.

It should be evident that maintaining a productive relationship and clear and regular lines of communication with a project steering committee is vital to ensuring the long-term success of a project. While we will get into communications and stakeholder management in a little more detail later in this chapter, here are basic items:

- Holding regular check-ins with the project steering committee.
- Focusing on their key project goals and concerns; these would be the benefits of the project for their own strategic objectives.
- Preparing for committee meetings:
  - agenda (if you set it);
  - effective and focused summary presentations of progress and issues;
  - ready for critical questions that may arise;
  - ready to listen and take careful meeting notes.
- Following up with committee members with summary meeting notes and addressing any questions or requests for information not provided.

At times you may not have an established project steering committee and may want to create one. If you have influential project sponsors and an engaged client, it can be valuable to create a project steering committee that comprises senior managers of your own organization and the client organization. These can separate groups, but I have found when the opportunity exists to combine the two, it is advantageous to bring the two sides together. This can help to foster transparency between the entities, promote alignment around the project goal and objectives, and address critical issues as they arise. This can also help to avoid any miscommunication around the project's scope and required resources, and competing agendas.

PROJECT STATUS MEETINGS AND AGENDA

The one thing that usually ranks at the top of complaints around corporate work is ineffective meetings. Project meetings can be a highly effective, valuable activity, or they can be the biggest drain on time if mismanaged. Just consider for a moment the number of team members, hourly rate each team member costs, and the time spent in a meeting, and how much value you are asking for when you conduct a meeting. You might consider whether it may be more advantageous to send an email, issue a summary report, etc. On the other hand, maintaining communication flow and checking in with the team collectively on project progress is instrumental, if done effectively.

There needs to be a clear purpose for meetings, only essential stakeholders involved, and, in addition to the agenda, set rules for how the meeting is conducted. The purpose of a project meeting (team, steering committee, etc.) is to share information on project status, discuss ideas, opportunities, and risks, and to come to some decisions or to schedule additional meetings when time is constrained and you need additional input from other stakeholders or only need a small subset from the group in the meeting.

When deciding who needs to be in a meeting, I suggest using the "Lifeboat Rule" by asking some simple questions:

- Who absolutely needs to be in the boat (meeting)?
- What do they bring to the meeting?
- If they are not included, what may be lost from the meeting?

Again, using the RAM and org. charts covered previously can help in deciding who needs to be in a specific meeting at the start of a project.

Project meetings need to be managed, like anything else on a project. Projects usually have several standard meetings during the project lifecycle:

- project kick-off meeting held at the project start
- regular project status meetings held periodically
- milestone or stage gate reviews at key project junctures
- project closure and lessons learned held at the end of the project

On agile projects there are set meetings that are part of the agile development process: sprint planning, daily scrums, client retrospective, etc., and each has its respective participants and formats. As this book is focused on standard project management, we are not going into detail on agile project meetings.

Below is an example of an agenda for a project kick-off meeting:

Project Kick-off for the Community Fire Resilience Toolkit
(in short the Fire Resilience Toolkit)
Agenda
3:00–4:30PM, Monday, May 6, 20XX
451 Firefighter's Path
Golden City, CA

Note: a separate and complete list of attendees will be included in a follow-up to this meeting.

Meeting Goal: Initiate the development of the Community Fire Resilience Toolkit.

Meeting Agenda/Objectives:

1. Introduce all team members and identify key stakeholders;
2. Review the project scope: goal, deliverables, constraints, assumptions/risks, and quality expectations;
3. Review the schedule: development process, major project milestones, status meetings;
4. Review risks and opportunities and measures to address;
5. Quick review of understanding of requirements, any current design ideas/previous work samples;
6. Next steps/meetings;
7. Other items, issues, questions.

Project status meetings should be brief and focus on key project elements. These meetings can be held as a "stand-up" meeting, quite literally done "standing up" and not sitting (unless a disability precludes this), common to ICS, and their intent is to cover basic task status and can be kept to around 15 minutes on a weekly or biweekly basis; a little longer if the team is larger than five team members. The project manager should lead with an overall report on project status – schedule and budget (if relevant), any changes to the scope, and client feedback (if available). Team members will focus on reporting out on their own task status– what was completed, what is being worked on, and what is coming up in the near term. Written agendas should include the following:

- Meeting goal/purpose
- Key items to address: scope, schedule, budget, and only other relevant project plan elements: risk, quality, procurement, etc.
- Who is taking part in the meeting
- Date, time, and location (even if held online; online systems such as MS Teams and Zoom provide a record of who attended, date and time, so they may be used for recording this info., but be mindful of legal regulations around workplace communication and electronic recordings)

Some agendas include timing for each item and who is leading each item on the agenda as well. Another option is to assign a note-taker (good opportunity for interns or junior PMs), a designated time keeper, and facilitator, often this the project manager, although these roles may rotate, which I recommend as opportunities for professional development of potential project managers.[4] Agenda and meeting details are usually driven by the organizational culture, so, if you are a new project manager, I recommend asking for a copy of a recent project meeting agenda of one for a senior level meeting to see what is standard. If there is none, then you develop and present your own standard for meetings.

Managing meetings effectively requires some skill and diplomacy to keep them on agenda, facilitate effective discussions, and arrive at decisions and actions. Here is a good set of rules that I recommend following to ensure effective meeting facilitation:

Before the Meeting

- When scheduling a meeting be mindful of the day of the week or time of day and normal business hours, as well as lunch breaks. Avoid meetings late in the day, just after lunch, or just before a weekend or holiday if you want to maintain attention and engagement.
- When there are items that need to be discussed in the meeting and require reading/ viewing, then make sure to send them ahead of time ("read ahead" materials) and make it clear that this is necessary and will be covered in the meeting, but make sure to hit key points in the meeting for the unprepared!

During the Meeting

- Make sure to register who has attended the meeting and any critical absences (this can be a warning sign of potential root causes for communication-based risks later on in the project).
- Review and track the agenda, and if this is a new meeting or there are new members (even if it is one new person), and ask if there is anything to add to the agenda. Mention if there are any special items or presentations that are planned in the meeting.
- Ask that participants check their job titles at the door to increase transparency of sharing their ideas and concerns. This is often difficult when there is a rigid organizational hierarchy or established managerial culture, but what has proven effective for increasing meeting productivity is a level of engagement and meaningful discussion around critical issues.[5]
- Make sure to engage everyone in the meeting to the extent possible. Some people are introverts and may be quiet. Be mindful that some people may be uncomfortable speaking in a meeting, so you may want to reach out to them beforehand and ask them to prepare a few thoughts on an important discussion topic, or to engage them with some questions they are prepared to answer to in the meeting.
- Follow the ELMO rule. ELMO stands for: Enough, Let's Move On. There will be times when a meeting participant goes on at length or two or three individuals discuss something for far longer than needed; or a specific issue requires more time outside of the meeting and fewer people to hash things out or other people not in the meeting. That is when you need to pull the ELMO card and politely interrupt the speaker(s) and ask that the item be addressed in a separate meeting. Some organizations go an extra step by using a set of props to manage meetings: flags, talking sticks, timers. etc.; I find that these are not necessary as long as everyone acknowledges the meeting format.
- Be mindful of the time, not only how long the meeting is running and if you are bumping up against the time limit set for the meeting.
- Close the meeting with any major decisions made, actions to take, next steps, and any meetings that need to be scheduled, upcoming meetings, and milestones. Thank everyone for their time.

After the Meeting

A critical item is making sure to follow up on meetings with the next steps, planned actions or decisions that have been taken. This means writing up a summary of notes and those key decisions and actions. This should include the following:

- The original agenda, who attended the meeting, date, location.
- Key issues discussed (status, risks, opportunities) and decisions made.
- Assigning who is responsible for specific next steps – actions or a decision – and the expected completion date.

Make sure to assign these tasks to all who were involved, and this may include senior people in the organization. Follow up and track status on these tasks as part of your regular log of project work. New project managers may be uncomfortable with the idea of writing out an assignment for a more senior manager to themselves. But remember, if they agreed to take responsibility for getting something done or making a decision, you are just holding them accountable for what they agreed to on the project.

NOTES

1 Henry Gantt and Frederick Taylor both worked together at Midvale Steel Works and at Bethlehem Steel, where Taylor developed his scientific approach to management, commonly called Taylorism, focused on time and motion studies to improve production efficiencies. While Taylor developed systems to set rates of production and pay per hour, Gantt developed a task and bonus system which paid bonuses over standard rates of production. This difference in management philosophy leads to Gantt parting ways with Taylor.
2 Tuckman, Bruce W (1965). "Developmental sequence in small groups". Psychological Bulletin. 63 (6): 384–399. doi:10.1037/h0022100. PMID 1431407
3 Katzenbach, J. R., & Smith, D. K. (2015). *The Wisdom of Teams: Creating the high-performance organization*. Harvard Business Review Press.
4 When I first managed a project, one of the informative books I read at the time was *Managing the Non-Profit Organization: Principles and Practices* written by the legendary Peter Drucker. One of practices I picked up and implemented from this was to rotate who led the meeting to keep the meeting fresh and to allow each team member an opportunity to practice leading a meeting. I found this to be a great learning experience, not only for the team, but also for myself, to see what my team can do and to learn from their practices.
5 Senor, D., & Singer, S. (2011). *Start-up nation the story of Israel's economic miracle*. Twelve.

Chapter 9

Developing the Quality Plan

Quality is not an act, it is a habit.

W. Edwards Deming, Engineer, Author, Management Consultant

QUALITY MANAGEMENT

Emergency and public safety professionals don't often dwell on the topic of quality. It's not that quality is not important; we often assume that based on our training, continuous improvement cycles, after action reviews, and attention to critical performance standards (NFPA, EMAP, etc.), and how critical public safety is, that it's inherent in the work we do. While this may be true, what is different when it comes to projects performed for public safety and emergency preparedness is that we are looking at quality from several different points of view.

Refer to Figure 9.1. We look at the quality of how the project is managed, in particular the project processes, and that includes much of what we have covered so far: developing the scope, schedule, budget and so forth, and how well they are managed throughout the lifecycle of the project. The next level down from this is the quality of the development work which culminates in the final deliverable, set of deliverables or outcomes achieved, and a product/service. One level of quality management affects the other; our project management processes influence the development work, and the work impacts the quality of the deliverables and vice versa.

WHY IS QUALITY IMPORTANT?

What project quality management brings to the table is a formalized quality management process with a number of different methods and tools. The quality plan is another element of the overall project plan that helps to define the quality requirements for the project, translated from the expressed client needs, and creates a framework to deliver an end result that satisfies those requirements. You can compare project quality management as similar to defining success in a game, marking the field of play and the goal posts, so it is objectively clear when the goal has been achieved.

DOI: 10.1201/9781003201557-11

Figure 9.1 Project quality Bullseye diagram.

DEFINING QUALITY

Let's come back to what we started off in this section, with an essential question: What is quality?

There have been many quality gurus over the years who have defined the term quality.[1] When the question is posed the answer often heard is "I can tell you when I see it", which we can translate into performance that the end-user can see, feel, and experience. With that in mind, we are going to use a very simple quality definition attributed to Joseph M. Juran: Fitness for Purpose.[2] Defining quality in terms of "Fitness for Purpose" means that whatever the product or service that is created, it needs to fit its intended purpose for which is was originally designed. This can be further broken down into the elements that form a product/service (for brevity I will use the term product to mean both product and service in this section):

- **Design and function** – what are the key functions that the product performs? What are the design features that enable the product to be used effectively?
- **Reliability** – how long and frequently can the product be relied upon to perform the intended function and maintain the design features over a specific period of time.
- **Durability** – how well can the product perform under certain conditions (specified) without declining or breaking.
- **Service and maintenance requirements** – how frequently and to what extent does the product require regular service and maintenance to support its performance.

All of these product attributes can be understood by the quality of their performance requirements. A couple of items to note:

- Both reliability and durability are based on the conditions under which the product performs. We will address this a little later on, as extreme events or missing (often assumed) conditions can influence performance qualities.

- I will be using the term quality requirements or just the term requirements to mean quality performance requirements.

QUALITY PLANNING

Now let's address the process by which we develop and manage quality requirements and other elements of the quality plan. To illustrate these quality methods and tools, I will use the software application project referenced in the previous chapter to walk through each of them. To provide more context for this example, here are the following parameters:

- Goal of the project is to develop a software application to manage registrations for special medical needs[3] client records in an emergency shelter for Liberty City, a mid-size city (300,000 residents), and to complete and deploy this within one year.
- Current place and time the project is set in is the Western United States in 2022, in the fictional town of Liberty City.
- Project is being developed by a software development company for the Liberty City Emergency Management Agency.

To elaborate this further, here are the major project deliverables[4]:

System Deliverables

- Client screening and registration
- Client medical triage and intake
- Client medication tracking (storage, handling, administration)
- Service animal/pet intake registration and care
- Client medical record wristband management
- Client bed assignment
- Client medical care and critical event management
- Client transfer to acute medical care
- Client discharge

System Support Deliverables

- Software manual and training
- Server stack required to host the application
- Software security and UX testing plan

PRELIMINARY PROJECT QUALITY STATEMENT AND POLICIES

The first part of a Project Quality Plan should include the following basic elements, a brief example is provided in Table 9.1 for the current sample project as given above.

Table 9.1 High-Level Quality Plan

Project quality definition	Quality is the performance capability of the software to allow for efficient, accurate, and secure screening, registration, and management of records for shelter clients with special medical needs during their stay at an emergency shelter.
Project client	Tina Gonzalez, Director of the Liberty City Emergency Management Agency (LCEMA)
Project end-user	Liberty City Memorial Hospital, American Red Cross volunteers, Medical Reserve Corps, other designated shelter staff assigned to the special medical needs branch
Critical quality requirements and strategy	The top requirements for this application are: • Data security and confidentiality (including HIPAA compliance) • Maintaining integrity of data • Availability of the application (uptime) • User experience will allow for easy client data input, client record search, and reporting • Software will be accessible for people with disabilities; Section 508[7] compliant • System must be interoperable with other shelter software applications and, with adherence to security restrictions, allow sharing client data with trusted partners Quality guidelines and methods will use a waterfall development approach and three levels of software testing: Unit Testing, System Testing, User Acceptance Testing, Performance Testing (using testing data), and Security Testing.
Quality policies, standards, and compliance	• HIPAA compliance • NIST Security Standards • Compliant with Liberty City IT Standards • ITIL Software Management Standards • Section 508 compliance

The first part provides a basic definition of what quality means for this project, making it project-specific, similar to the project goal. This is useful as the project proceeds through development and can be used by the project team and key stakeholders to ask the foundational question, "Does the project quality meet the performance capability...". The second parts define the Project Client and Project End-user, as illustrated in Table 9.2. These bits of information may be included in the scope document at the beginning of the project plan, but it helps to keep the focus on the customer.

We differentiate between these two customer categories as we need to balance the quality performance expectations for these two sets of stakeholders as there may be differences. For example, the Director of LCEMA may put greater emphasis on legal compliance to satisfy the City's Law Dept. and the IT Dept's NIST standards since they may be a "Show Stopper" for the project (i.e. may call a halt to or end the project), whereas the end-users are looking for ease of use and sharing data with outside partners. The Critical Quality Requirements and Strategy defines high-level requirements and how they will be managed during the project life-cycle. This is followed by the Quality Policies, Standards, and Compliance Requirements which identifies specific guidelines and performance standards that the project and end-product needs to meet.

DEFINING PROJECT QUALITY REQUIREMENTS, REQUIREMENTS MANAGEMENT, AND REPORTING

Referring to Table 9.2, the next major step in developing the quality plan is defining the requirements in SMART terms based on an initial listing of needs, what is called a *Needs→Requirements Matrix*. This starts with a general list of what the end-user

Table 9.2 Quality Requirements and Management Plan

Client Needs from Product/Service→	Critical SMART Requirements
Screening clients accurately	Performing at 99.99% accuracy in screening clients according to special medical needs criteria.
Registering shelter clients accurately	Recording shelter client data consistent with registration forms at 99.99% accuracy
Efficiently manage client records	• Average time to record updates to patient records is within the normal range of similar applications (avg. time > 31 s). • System accessibility measured by 99.99% uptime.
Efficiently share client records with approved partners	• Recording and managing user accounts for approved partners (avg. time > 61 s for entering new partner account; update avg. > 31 s). • Recording and providing limited access to client records tracking usage by partners, including client records shared.
Keeping client records secure	• Meets HIPPA requirements including client record privacy, security, and breach monitoring and signal for security notifications. • Meets Info. System security requirements stated by Liberty City Dept. of Information Systems (LCDIS) including NIST Standards
Requirements management Methodology	Requirements for the applications will be identified using three key quality methods: • Process mapping with application end-users (using existing registration and special medical needs shelter records). • Review of better practices with similar applications. • Fault tree analysis based on previous experience with records management. • Rapid prototyping within the waterfall development process and testing at key development stages (as given above). During the application development lifecycle, a quality register will be used to manage quality requirements and specifications, including acceptance criteria, validation, and verification processes.

Requirements acceptance Methodology	**Key Requirements and validation process, acceptance criteria, and role approving acceptance.**	
	Key requirements	• Data security and confidentiality (including HIPAA compliance) • Maintaining integrity of data
	Validation process	• Verification of HIPAA compliance checklist • Security and unit testing, system testing
	Acceptance criteria	Passes all tests
	Approver	Director of InfoSec, LCDIS Director of LCEMA
	Key requirement	Availability of the application (uptime)
	Validation process	Performance testing (using testing data)
	Acceptance criteria	System accessibility measured by 99.99% uptime
	Approver	Director of Application Development, LCDIS Director of LCEMA
	Key requirement	User experience will allow for easy client data input, client record search, and reporting
	Validation process	User acceptance testing
	Acceptance criteria	• Average time to record updates to patient records is within the normal range of similar applications (avg. time > 31 s). • End-user feedback survey.
	Approver	Director of Application Development, LCDIS Director of LCEMA
	Key requirement	Software will be Section 508 compliant
	Validation process	Section 508 Compliance Test by Liberty City University's Program on Disabilities Studies, Web Accessibility Lab
	Acceptance criteria	Web Content Accessibility Guidelines (WCAG) 2.0
	Approver	Director of Application Development, LCDIS Liberty City Office for People with Disabilities and Access, and Functional Needs (LCOP-DAFN)

(Continued)

Table 9.2 (Cont.)

Client Needs from Product/Service→		Critical SMART Requirements
	Key requirement	System must be interoperable with other shelter software applications and allow sharing client data with trusted partners
	Process	Unit testing of middleware data exchange and data exports and imports for other applications
	Acceptance criteria	99% Data integrity and usability for managing client records after transfer
	Approver	Director of LCEMA American Red Cross Medical Reserve Corp.
Project reviews and assessments Methodology	Review type	Project quality review – Internal
	Frequency	Biweekly
	Reviewers	Project Manager, Project Team
	Reports	Punchlist of quality issues and improvement plan (Kanban tracking)
	Review type	Project quality review – Overall/External
	Frequency	Once a month
	Reviewers	Project Manager, Director of LCEMA
	Reports	Corrective Action Plan (Kanban tracking)
	Review type	Phased Product Quality Plan: QA/QC Testing
	Frequency	By development phase: prototype, version 0.1, version 0.2 until Final Version
	Reviewers	Project Manager, Quality Manager, and QA/QC team
	Reports	Application Bug Reports (Bug Tracking Report)
	Review type	End-user testing
	Frequency	End of development phases by versions
	Reviewers	End-users, Director of Application Development, LCDIS Director of LCEMA
	Reports	Bug Fixes, Final Approval (Bug Tracking Report)
	Review type	Security Testing
	Frequency	Final Version
	Reviewers	Director of Application Development, LCDIS Application Security Specialist
	Reports	Security Fixes, Final Approval
Tools		• 3 servers: development, testing, and production servers with software configuration for application • End-user configured desktops with WiFi and all internet web browsers for testing • Electronic Kanban (planning board; Jira, Trello, Miro, Asana, etc.) • Bug tracking software (Redmine, trac, etc.) • Share document – Quality register in MS Excel
Communications		Following the quality plan, the overall project plan will be reviewed where quality assurance may be applied, e.g. scope management, communications, and risk. Any quality issues will be logged in the Kanban and updated in the other plans on a regular basis.

(customer/client) needs from the product and can be stated in general terms and do not have to be described in measurable terms. Needs can be identified based on the scope and set of deliverables.

In Table 9.2, one need for the application is "Screening clients accurately". This is followed by the next step which translates needs into requirements. In the same example, we state requirements in SMART terms; hence, the related performance requirement is: performing at 99.99% accuracy in screening clients according to special medical needs criteria.

We also describe the Requirements Management Methodology which explains how quality requirements will be identified and managed on a more detailed and methodical basis. In the example below, the project uses process mapping, a review of better practices, Fault Tree Analysis, and rapid prototyping within the waterfall development process to elaborate the requirements and manage them. This is followed by detailing the Requirements Acceptance Methodology, how Key Requirements will be validated, defining the acceptance criteria, and who will approve the acceptance that requirements have been met.

The next section includes a list of regular quality review meetings, who, what, and when they are reported on. The final sections explain what tools are required, in this case systems and applications, and the last item is how quality will be communicated and connected to other planning processes and documents (scope, schedule, risk, procurement, etc.).

The high-level plan is intended to provide a blueprint to guide and manage project quality during the design and execution phases. The next major steps are identifying detailed quality requirements and developing procedures for quality assurance and quality control, including who is responsible, frequency, and the quality data collection process and reporting format.

DEFINING QUALITY REQUIREMENTS: IDENTIFICATION TOOLS

Defining the key questions that concern quality for project deliverables begins with a discussion with the key stakeholders: the product end-user group, client team (may be separate from the end-user group or may be the same), and the project champion (can be internal to the organization or a connected to the client team). For example, with the software application project, Tina Gonzalez, the Director of LCEMA, would be the Project Champion and, along with the representatives of LCDIS and LCOP-DAFN, might comprise the Client Team. The end-user group may comprise staff assigned to the Special Medical Needs Shelter from Liberty City Memorial Hospital, American Red Cross volunteers, Medical Reserve Corps, and other designated agencies.

When convening groups to identify product quality requirements, it is best to meet separately with each group, as the perception of project requirements will likely differ. It helps to manage the influence senior executives may have over more junior staff members in these type of discussions in order to avoid ignoring critical quality requirements from surfacing. Once all quality requirements have been identified, then their importance can be weighed based on these groups and client priorities, using a priority matrix we will cover later in this section on quality. Priority should be given to end-user quality requirements vs. those who are farther from the use of the product.

One simple method that can be effective in establishing preliminary quality requirements is to use the *Needs→Requirements Matrix* illustrated in the example above. This may be done through client interviews, a client focus group, or surveys or a combination of the two. The most productive method that I have found is to meet with the client or customer (end-user) at their worksite, discuss the product requirements and then observe and record the process in action. This may include exercises of all types (FSE, FE, TTX, drills, etc.), a hands-on walk-through of emergency response procedures with practitioners, or an in-depth review of an After Action Review/Improvement Plan (AAR/IP). When looking at the

AAR/IP, the individual improvements may be mapped to the FEMA Core Capabilities, as discussed in Chapter 3, or similar capabilities frameworks.

Let's revisit the concept of setting SMART objectives, as this also applies to quality requirements. We covered this in Chapter 5 under scope, but for easy reference quality requirements, to the extent possible, need to be specific, measurable, attainable, relevant, and time-bound. Meeting these criteria helps to better define potential product functionality and features. Again, the Core Capabilities and similar guidelines can be helpful as a starting point in defining them.

Additional areas to review for collecting quality requirements:

Historical (Looking at the Past)	Current/Future Looking
• AAR/IP/Hotwash Feedback (Lessons Learned) • Client/customer calls to a Call Center (may include text chat) • Complaints from clients/customers (may also be end-users) • Contract issues from past projects/services • Software bug tickets and reports • Social media posts • Data analytics for past hazards	• Client/customer interviews • Observation of processes and procedures • Benchmarking capabilities for similar agencies in other jurisdictions or other agencies/organizations • Data analytics for hazard simulations

Some specific methods for identifying quality requirements:

• Brainstorming with clients/customer/end-users
• Interviews with subject matter experts (SME) or clients/customer/end-users
• Surveys with client/customer/SMEs
• Conducting focus groups
• Prototyping and product demonstrations
• Process mapping
• Fishbone (Ishikawa) Diagram
• Six Sigma DMAIC Process: define, measure, analyze, improve, and control[5]

These methods may be used in tandem with one another, more suitably sequentially to help develop an initial set of quality requirements and then elaborate them further. For example, a project team developing a new set of standard operating procedures for emergency commodity distribution might work with the client team (including end-users) to conduct a brainstorming session to establish the foundational quality requirements, and then follow up with specific teams, e.g. logistics, mass care, legal, and use a Fishbone Diagram on specific areas of the foundational quality requirements.

Referring to the template in Figure 9.2, the Fishbone Diagram (also called an Ishikawa Diagram) is a type of root cause analysis (which can be used for both identifying quality requirements and groups quality problems into four major areas, sometimes called the 4Ms: people [what used to be referred to as "Man"], methods, materials/supplies, machines/measurements). An example is given in Figure 9.3 of a simplified Fishbone Diagram that might be conducted with the Mass Care Team, drawing on their experience running a Commodity Distribution Point (CDP; for emergency incidents) with no standard operating procedures.

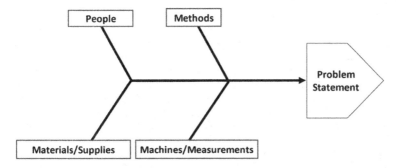

Figure 9.2 Fishbone (Ishikawa) diagram model.

Referring to Figure 9.2, this starts with an overall problem statement (does not have to be SMART, but aspects of the model can help) and then breaking this down into the categories and asking the question "Why" five times to get at the root of the problem. Categories are explained as follow:

- **People:** people involved in the service or product delivery – staff, employees, vendors, contractors
- **Methods**: the methods used to deliver the service/product. This can include procedures, steps, checklists.
- **Material/Supplies:** materials or supplies (mostly consumables but can be durable) used in the process.
- **Machines/Measurements:** machines and measurements used for service/product delivery. Measurements are included as they are often part of a production/service process. This may include dashboards etc.

To follow on with this example, we start with the problem statement:

❶ Emergency commodity distribution in the past has been problematic due to poor communication with residents and difficulty in managing resident expectations. This has resulted in complaints by residents and community political leaders, poor allocation of food commodities, and inequitable distribution (more vulnerable communities underserved).

We then ask the question "Why?" and begin listing these problems according to their area as part of the process. Continuing to ask why until we get to the root of the problem. If we proceed at the top under method, we identify one of the first reasons that communication with residents may pose a problem:

❷ Are not able to communicate with residents who have limited English speaking skills

If we ask "why" again, we may find:

❸ Staff do not have access to language interpretation/ translation

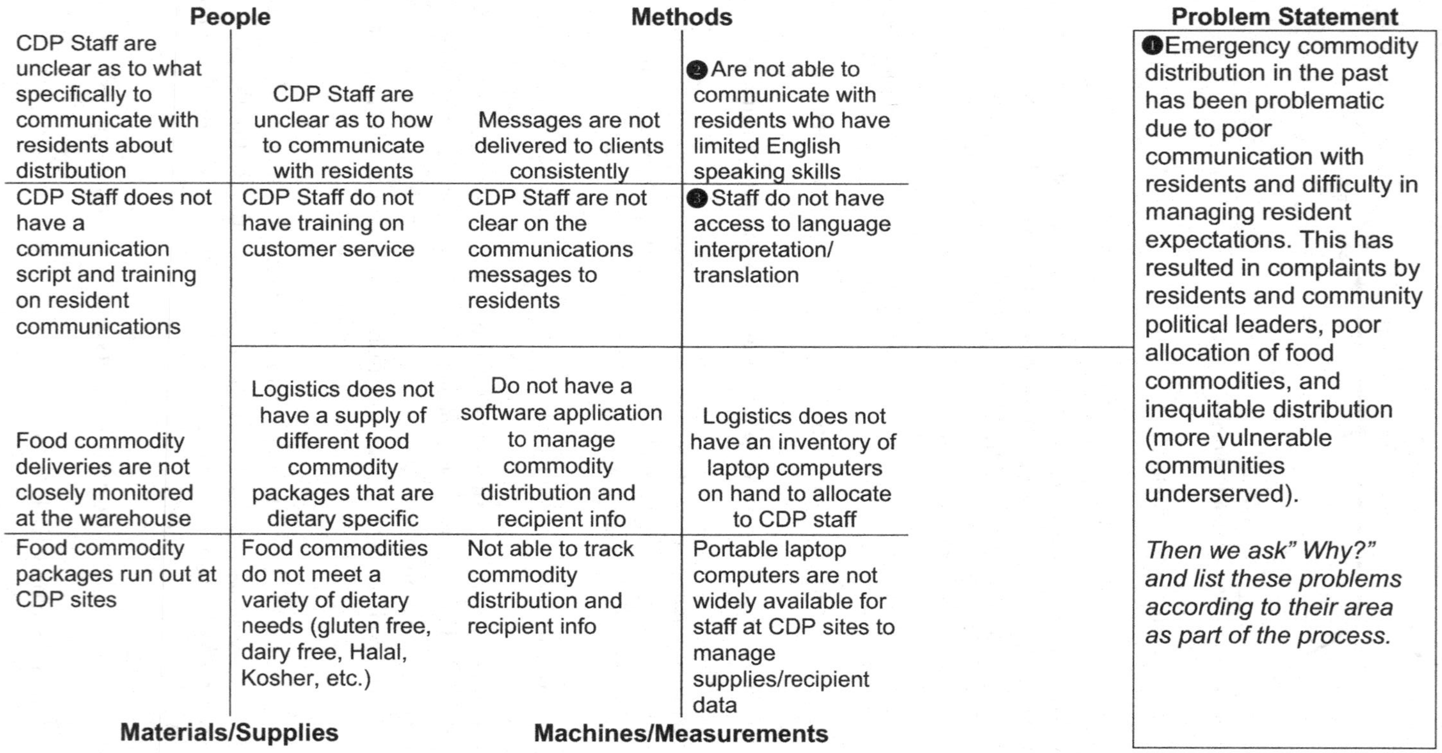

Problem Statement

❶ Emergency commodity distribution in the past has been problematic due to poor communication with residents and difficulty in managing resident expectations. This has resulted in complaints by residents and community political leaders, poor allocation of food commodities, and inequitable distribution (more vulnerable communities underserved).

Then we ask" Why?" and list these problems according to their area as part of the process.

People

- CDP Staff are unclear as to what specifically to communicate with residents about distribution
- CDP Staff does not have a communication script and training on resident communications
- CDP Staff are unclear as to how to communicate with residents
- CDP Staff do not have training on customer service
- Messages are not delivered to clients consistently
- CDP Staff are not clear on the communications messages to residents

Methods

- ❷ Are not able to communicate with residents who have limited English speaking skills
- ❸ Staff do not have access to language interpretation/ translation
- Logistics does not have a supply of different food commodity packages that are dietary specific
- Do not have a software application to manage commodity distribution and recipient info
- Logistics does not have an inventory of laptop computers on hand to allocate to CDP staff

Materials/Supplies

- Food commodity deliveries are not closely monitored at the warehouse
- Food commodity packages run out at CDP sites
- Food commodities do not meet a variety of dietary needs (gluten free, dairy free, Halal, Kosher, etc.)

Machines/Measurements

- Not able to track commodity distribution and recipient info
- Portable laptop computers are not widely available for staff at CDP sites to manage supplies/recipient data

Figure 9.3 Fishbone ((Ishikawa) diagram for identifying quality issues with a commodity distribution point.

We can continue to go through this process, identifying further root causes for each part of this problem, for example "Why do staff not have access to language interpretation/translation?" And then come to the answer: the agency does not have access to a language interpretation or translation since there is no translation service or language interpreters who have been hired.

Each of the factors that contribute to the main problem statement can then be used to develop product requirements. If we are developing the standard operating procedures for emergency commodity distribution, one of these would include a procedure that provides language interpretation and translation. Figure 9.3 provides an elaborated Fishbone Diagram for the above reference project.

As discussed earlier in this section, project quality requirements should be collected (in a quality requirements register; covered later) and sorted by priority according to what is critical to quality. This is an art and science to balance the level of influence and needs for the major client groups: sponsor (point of contact, usually an executive manager), client team, and end-users. In some cases, they may be part of the same group, but quite often with larger government-funded development projects in emergency and public safety management, they are related but at separate managerial levels. With this in mind, it is helpful to use a priority matrix to evaluate and weight the level of importance for each set of requirements by deliverable. This prioritization can then be used to discuss and confirm these priorities with the collective client team. We will address the method and tools for a prioritization matrix when we discuss project portfolio management in Chapter 11.

QUALITY ASSURANCE PROCESS

The purpose of the quality assurance process is to develop the set of procedures to ensure that quality is managed at the process level on the project, as opposed to quality control which is designed to manage the quality of outcomes (this may be an interim process inspection step or end-product or service). One of the best ways to illustrate the difference between quality assurance and quality control is provided in Table 9.3, a SIPOC diagram, using the example above for CDP with the requirement that a food package supplier and distribution process meets a variety of dietary needs.

Most emphasis in quality planning is placed on quality assurance as we want to "design quality into the process". This can be viewed from the project process level, which is following better practices in project management as laid out in this book, at the product development level, or at the operational level for any product or service that will have ongoing management, maintenance, and support, e.g. software application and plans. Some examples of quality processes in the development cycle can be the quality managed in the software development process; design and development methods and tools used for building levees; or development of new surveillance procedures for bodycams used for public safety.

The most common, standard set of measurements that can be easily applied to emergency and public safety related projects are the FEMA Core Capabilities and the National Incident Management System (NIMS) typology for resources. These standards provide a clear set of specifications of performance and are integrated with one another. They also provide a clear SMART format, discussed earlier in the book, for stating these capabilities.

Table 9.3 SIPOC Model for CDP Food Packaging

Supplier	Inputs	Process	Outputs	Customers
Food packaging wholesaler	Food commodities	Distribution of food packages at a CDP	Food packages targeted to specific dietary requirements	Residents receiving food packages

Quality Requirements	Quality Measurements	Quality Requirements	Quality Measurements	
Packaged food commodities that are dietary specific	• Food packages with verified food sources for dietary needs (Halal, Kosher, dairy free, and nut free) • Process inspected by certification companies twice a month • 100% food packages marked as meeting dietary needs	• Communication pieces targeted to residents on qualified food certification for dietary needs • Capability to sort, store, and distribute food commodities by dietary needs • Resident able to identify food packages by their dietary needs and trust the certification	• 100% of packages are marked with food certification and ingredients • Each food package has a color code on the outside for easy identification and separate handling • Flyers are printed with the food certification process and verified certification, including multiple languages • Joint inspections by community representatives and agency of food suppliers conducted twice a year to verify supply and certification process	

For example, a public safety agency wants to develop one of the core capabilities under *On Scene Security, Protection, and Law Enforcement*: "provide and maintain on-scene security and meet the protection needs of the affected population over a geographically dispersed area while eliminating or mitigating the risk of further damage to persons, property, and the environment"; the following is one of the SMART capabilities that can be adapted for a law enforcement agency:

> Within (#) (time) of an active shooter incident, provide security and law enforcement services to protect emergency responders and (#) people affected.

We can use this as a clear performance specification (capability) that we are looking to develop. Identifying appropriate resources that would be inputs (as part of a SIPOC analysis) are also included in this core capability:

- Patrol Team
- Mobile Field Force
- Special Weapons and Tactics Team

All these resources are included in the NIMS typology, along with recommended national level first responder training courses[6]:

- AWR-235: Initial Law Enforcement Response to Suicide Bombing Attacks, Mobile
- AWR-335: Response to Suspicious Behaviors and Items for Bombing Prevention VILT(Office for Bombing Prevention)

- PER-200: Field Force Operations
- PER-265: Law Enforcement Response Actions for CBRNE Incidents
- PER-340-2: Active Threat Integrated Response Course, Indirect

These trainings may be considered part of the specifications for having trained resources capable, along with any plans, equipment, and other materiel and supplies, for delivering this capability. In sum, with these two sets of specifications, we have defined who we need by the NIMS typology and how they would be trained according to national training standards.

The next step in the quality assurance process is to develop the quality assurance activities to verify that the specifications have been met. Activities for verifying specifications (capabilities) may include:

- Tabulating what has been completed: number of trained staff, amount of specific PPE items.
- Evaluating the quality of individual of team performance may be through drills. This may include recording or reviewing the performance with external evaluators.
- Use of surveys for assessing quality criteria, for example feedback on training.
- Reviewing the results of a process, for example load testing a software application for processing a range of user records.
- Conducting an inspection or test of specific item, for example the performance of a ventilator on screening out certain size particles.

In our example above, several quality assurance activities may include:

- Identifying the number of members of the *Special Weapons and Tactics Team* who have completed the training course *PER-265: Law Enforcement Response Actions for CBRNE Incidents*.
- Evaluation in a drill of a *Patrol Team* in creating a secure perimeter (specifications would be determined) after a simulated active shooter incident.
- Using a validated survey to assess knowledge transfer for a specific training as part of response team competency building.

QUALITY CONTROL PROCESS

The bottom line for any public safety agency or emergency management organization (dept., office, etc.) is how they perform on the general mission of life-safety, protecting infrastructure and property, and mitigating the causes and impacts of an emergency incident. The benefits from a project in public safety or emergency management are measured along similar lines, and it needs to address the foundational question: how much will the project help enhance carrying out that mission?

So far in the quality management process we have covered how to identify quality requirements and develop activities for quality assurance, the latter focused on project development processes. The purpose of the quality control process for a project is to verify that a measurable outcome or deliverable has been met, in other words, end results. This may be to control for quality on the entire product of the project or a deliverable of the project. If we look back at the SIPOC diagram presented earlier, it is the Outcome of the process that is delivered to the Customer (s) that we are looking to measure. However, some development processes may have a number of chains of the SIPOC, so we are looking at the end of the chain.

QUALITY CONTROL PROCESS METHODS

The Quality Control Process seeks to check project development outcomes, deliverables/objectives, by either validating performance or verifying it against measurable specifications, and checking whether guidelines, processes, and procedures are adhered to, and identifying any deviations from the project planned quality to analyze the results and provide corrective actions and improvements. The following are components of quality control:

- **Validation:** evaluation of whether the end result (product/service) meets the performance expectations of the customer.
- **Verification:** evaluation of whether the end result meets a defined specification.
- **Monitoring:** assesses whether guidelines, processes, and procedures are being followed according to plan, collecting information and data.
- **Analysis:** analyzing quality data and developing reports and identifying deviations from performance expectations and specifications for potential areas of improvement.

QUALITY VALIDATION AND VERIFICATION

Validation is a quality control process of looking at what performance requirements the client/customer expects, in other words can observe, experience, and provide feedback on to validate that they have been met. The most common forms of validation methods are product demonstrations or walk-throughs of product functions and features. Another example of this is testing with prototypes at an early development stage, which may sometimes involve the use of mock-ups or wire frames (for software User interface design) to approximate the user experience. Validation methods may include user feedback forms or surveys, using both quantitative and qualitative measures linked to performance requirements. Using these forms can help reduce subjective opinion.

Verification, on the other hand, is a quality control process that evaluates whether a measurable specification has been met. This an objective process, and the most common forms are measuring physical or observable dimensions of the product, its measurable performance within certain tolerances. This may include tests aligned with product or deliverable specifications.

Using the project example of the software application for client registration at a Special Medical Needs Shelter, this might look like the following for the requirements for user experience in recording medical records:

Validation Process for User Experience (UX)	Verification Process for UX
Conduct survey of feedback from a demo session with ten diverse end-users on the ease of entering recording client medical records (using dummy data) over 20 min. Survey will include quantitative (Likert scale) and qualitative open feedback.	User testing (ten data entry staff), entering client records (dummy data) over 10 min to measure whether the average time to update record patient records is within the normal range of similar applications (avg. time > 3 min).

MONITORING QUALITY CONTROL

Monitoring quality control is a higher level activity that assesses whether guidelines, processes, and procedures are being followed according to plan. This functions similar to an audit, except it is usually performed by the project manager or team members. Monitoring quality control may include reviewing standard quality operating procedures, technical documents, or an inspection of the development process as it is being performed. This would be done according a schedule, frequency may depend on the cycle of production, how critical that item is, and the practicality of controlling a specific area of quality requirements.

Using the User Experience (UX) example above the Project Manager would review the UX quality validation process by assessing the demo session: the survey development process, the number of end-users, look at the diversity of the end-users, and minutes the session was conducted; all to determine whether this element of the quality validation process was conducted according to plan. The end result of the quality monitoring process would be a report on how well the quality control process is being followed.

One tool commonly used to record and track quality requirements is a quality register. See the example in Table 9.4.

QUALITY ANALYSIS

The quality analysis stage includes analyzing quality data and reports and identifying deviations from performance expectations and specifications with the goal of identifying corrective actions and areas of improvement. The most common tools for representing data are tables, charts, and diagrams to visually display and highlight that quality specifications are on or off target.

Among the basic data recording tools is a Check Sheet, which may be derived from a checklist of specifications or requirements, or also from user experiences with the product. Table 9.5 provides a simple illustration of a check-sheet for a pilot test of the delivery of an eLearning (online) course, using data from an IT helpdesk to identify any quality issues with end-users logging into a Learning Management System to take an eLearning course (an online system for managing and recording eLearning courses).

If we want to get a clearer picture of the data contained in the check sheet above, then we might display the data in bar chart form, such as a Pareto Chart. A Pareto Chart is a form of a bar chart for data that displays groups of data in rank order from largest

Table 9.4 Quality Requirements Register

No	Source (WBS, Client, Organization)	Requirement and Performance Specification /Category (SMART – Short Description; Deliverable, Operational)	Critical to Quality (High-Med-Low; 3–1)	Quality Assurance (QA) Activity and Measurement Activity	Quality Control (QC) Activity	Reporting Process and Metrics (QA, QC)	Responsible Team/ Staff Member	Schedule Frequency
1	Organization, WBS	Data security, integrity, and confidentiality (including HIPAA compliance), operational	3	Verification of HIPAA compliance checklist Security	Security unit testing and system penetration testing	QA – Report on HIPAA checklist QC – Report on security unit test, penetration testing report	Application Security Specialist – Rivka Braunstein	QA – After each unit of software has been completed QC – After each unit, when software beta version is ready, final release
2	Client, WBS	Uptime (application availability)	2	System accessibility measured by 99.99% uptime over 365 days	Performance testing (using testing data) in a simulated cycle of use and changing server environments (cloud, local based)	QA – Performance test report on uptime on Beta version in a simulated environment QC – Performance test report after release	Software Quality Assurance Specialist – Kevin Jackson	When software beta version is ready (simulated use), final release +6 months and +9 months
3	Client	Software will be Section 508 compliant	3	Software UI Test in the LC University Web Accessibility Lab	Section 508 Compliance Test by Liberty City University's Program on Disabilities Studies	QA/QC Compliance Test Checklist Report on Section 508	Software Testing Manager – Katherine Yen	UI test after each unit is complete, Beta and final release

Table 9.5 Check Sheet for IT Helpdesk

Date	Request Reset Password	Password Email Not Received	User Acct. Not in System	Course Stops	Course Does not Display	Other	Total
6/12/14	I						I
6/13/14	3	I	2				6
6/14/14	2		I				3
6/15/14	I		2	2		I	6
6/16/14	3		I	I			5
6/19/14	I				I		2
6/20/14	2		I	3		I	7
6/21/14	2		I	I			4
6/22/14	2		I				3
6/23/14	4					I	5
6/26/14	2						2
6/27/14	5	3	3	I	I		13
6/28/14	10	4		2			16
6/29/14							0
6/30/14	8			2	I	I	12
7/3/14		2	I				3
7/4/14							0
7/5/14						I	I
Total	**46**	**10**	**13**	**12**	**3**	**5**	**89**

number to smallest number of each grouped type of issue and then usually includes a line illustrating a running total sum of all issues encountered with the product.

Referring to the example given in Figure 9.4, let's suppose we had over 500 end-users as part of the eLearning pilot to test course delivery. Analyzing this date, we can see that Password Reset issues are more than 50% of all issues (46/89), so we would want to drill down into what is causing the password issues. The next problem is end-users not finding a user account in the system, so we would do the same, looking into the root causes for missing user accounts.

The following diagram examples are some other common data visualization charts and potential uses for quality analysis.

Refer to Figure 9.5. Another chart that may be used with a bar chart form is a histogram, sometimes called a Bell Curve, which shows the distribution of a data set for one variable over the range of potential outcomes. An example is given below for a histogram bar chart. A histogram may be useful in establishing the range of attributes required for a population, for example sizes of PPE (ventilator masks, hazmat suits), time to perform a building size-up, or size of generators commonly needed for backup power at specific emergency sites.

Refer to Figure 9.6. Scatter plot diagrams are helpful for charting two data variables against one another to analyze any correlative patterns between them. An example of its use for quality analysis may be looking at hours of training vs. measured performance in drills.

Refer to Figure 9.7. A Pie Chart takes multiple variables and charts them by percentage for each variable based on the total whole number from the data set. In the example given above we can see the percentage breakdown of source for firearms possessed by Federal

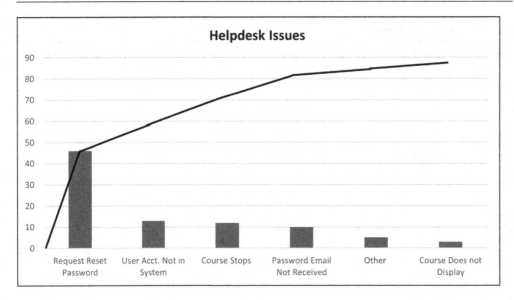

Figure 9.4 Pareto chart of helpdesk issues.

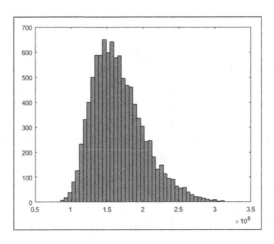

Figure 9.5 Histogram.(From Wikimedia Commons, L-W-Doyle.)

prison inmates. Pie Charts help to easily pinpoint the key quality areas to address for a given set of quality performance variables.

Refer to Figure 9.8. A Run Chart takes one data variable and charts them for each item produced or customer interaction according to a specific data point over time and includes control limits, upper and lower (UCL, LCL), indicating a range of acceptable performance, or an upper benchmark that needs to be met. An example is given below of a Run Chart.

A Run Chart is used to identify any deviations from the acceptable performance limits, and mostly used for tracking a fixed performance standard or one that has certain ranges of tolerance for performance. Some examples of this might be online application service uptimes, IT service levels, response times for EMS or building fire calls, and number of failures per a quantitative set (for example per thousand and per million).

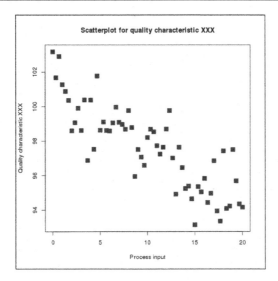

Figure 9.6 Scatterplot chart.

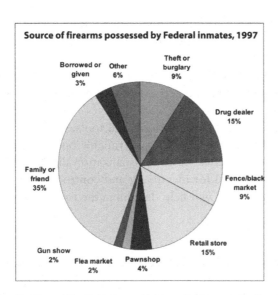

Figure 9.7 Pie chart. (Credit: Daniel Penfield. From Wikimedia Commons. See: www.ojp.usdoj.gov/bjs/
pub/pdf/ffo98.pdf. Chart created by User:AudeVivere.)

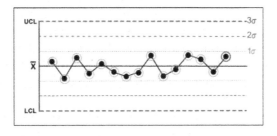

Figure 9.8 Data Run chart. (From Wikimedia Commons, GMcGlinn~commonswiki.)

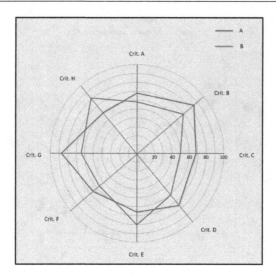

Figure 9.9 Spider diagram. (From Wikimedia Commons, Klever.)

Refer to Figure 9.9. Spider Diagrams, also called a Radar Chart, can be used for multiple data points that a product needs to meet and measure against the performance benchmarks. Using the example above, there are eight criteria given in the diagram (A–H), and if the inner line (for most points) indicates the defined product quality performance requirements and the outer line the actual product performance, then we can see that product does not meet criteria E and H.

There are other forms of data visualization and tools, some embedded in spreadsheet programs. A variety of software applications available on the commercial market offer a wide range of colorful and effective data visualization displays. Ultimately the quality data you collect, what is useful for identifying and controlling for quality requirements, and how that data is displayed will help determine the tools and type of quality reports you will use.

NON-QUANTITATIVE OPEN-ENDED QUALITY FEEDBACK

Some feedback from customers (clients, end-users) may come in the form of open-ended feedback. Working with open-ended feedback to perform a data analysis can be very challenging, so organizing the data first helps to then develop a comprehensive, thorough, and meaningful analysis. One method for grouping open-ended data is through affinity diagraming. Affinity diagramming sorts data by common groups and can be an iterative process, as some groups may contain a large number of feedback items that may need to be broken out into subgroups. Refer to Figure 9.10 for an affinity diagram assembled from open-ended feedback from survey forms concerning areas for improvement for a Logistics Center training.

We might take multiple comments and sort them into a common response type, and then group them into a common category. A second step might be to look at the frequency

Affinity Diagram: Feedback on Improvements to Logistics Center Training

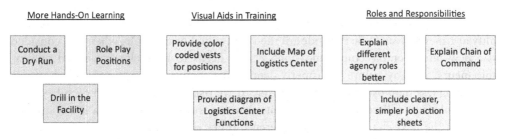

Figure 9.10 Affinity diagram.

of that feedback to then rank order the data using some of the previous data analysis tools we covered earlier.

QUALITY IMPROVEMENT

The quality improvement process looks at the deviations from planned performance or outcomes and identifies corrective actions or enhancements that can be taken for quality improvements. These are most often within the project development processes, but they may also involve adjustments in other project plan elements: the human resource plan, budget, schedule, procurement, etc. The process begins from the end of the quality monitoring and control process, with the outputs of the quality validation (customer requirements) and quality verification (specification measurements) processes. This analysis helps to identify what areas have been identified for improvement, how the development or production process needs to be improved, and how this can be tested and verified as complete. The quality register is a common, basic tool that helps to track major quality.

Some common tools used for quality improvement may include some of the same tools for verification as covered in the quality analysis process for monitoring and control, and additional tools:

- A process map or flow chart. A process map may also include functional or position swim lanes.
- Fishbone (Ishikawa) Diagram.
- Heatmap for different project attributes or stakeholders.
- Spider (radar) chart.

The end result of quality improvement is a report on what interventions were undertaken, what impact they had in making the expected improvements, and reporting this back to the key stakeholders involved, as well as identifying any further action required to maintain adherence to quality requirements. The monitoring and control to analysis and improvement is an ongoing process throughout the project execution phase until project completion.

PROJECT QUALITY VS. PROJECT RISK: TWO SIDES OF THE SAME COIN

Project quality management is very similar to project risk management. In fact, risk and quality at times appear to be two sides of the same coin. For example, when you consider that a new interoperable communications system needs to have clearly defined quality requirements, and any failure to properly detail those requirements may run a risk of the system being less than optimal. On the flipside, if we risk not meeting with each key project stakeholder to collect their requirements then we will likely result in a lower quality solution. In the next chapters, we will reference other related plan components, such as the risk, procurement, and stakeholders and communications, and will relate them back to quality assurance and control.

NOTES

1 Some of the more well-known quality gurus and their quality definitions: W. Edwards Deming defined "Good quality means a predictable degree of uniformity and dependability with a quality standard suited to the customer."; Philip B. Crosby defined it as "Conformance to defined requirements."; Gerald Weinberg defined it as "Value that the product provides to people."; Genichi Taguchi defined quality is the focus on product design to reduce losses to its intrinsic functions after being shipped. This is a "Loss Function" view of quality which focuses on product and production process design to eliminate variation and the potential loss in function.

2 Rose, K. (2014). Project Quality Management: Why, What and How. J. Ross Publishing. Joseph M. Juran wrote the *Quality Control Handbook*, which was first published in 1951, with a total of seven editions, the last published in 2010, shortly after Juran's death. Juran wrote originally that quality was defined as "Fitness for Use", changing this to "Fitness for Purpose" to address quality more broadly and to include service-based operations. He worked at the fabled Bell Labs in its hay day, taught quality at NYU, and was famous for the "Juran Trilogy" of Quality Planning, Quality Control, and Quality Improvement. Interestingly, the ideas of quality sampling of the end-product developed by Frederick Taylor (mentioned earlier in this chapter) predominated at Bell Labs. Juran's focus was on the human element and the quality processes.

3 Special Medical Needs in the sphere of emergency management have been defined as sub-acute medical needs that cannot be met independently in a general population shelter, but not sufficient to warrant hospitalization that may be met by basic skilled nursing and moderate levels of professional medical support, akin to an assisted living facility. It is intended to support for basic levels of care to reduce the pressure on hospitals and acute care providers during an emergency incident that requires mass sheltering care.

4 This is fictional example for illustration purposes. Most software development projects use Agile Project methods, so the basic software functionality might be mapped out in term of user "stories" (sometimes grouped into higher level Epics depending on the project size and agile development methodology employed) and then broken down into tasks. As stated at the start of this book, we are focusing on standard project management methods, while referencing where appropriate related project management methods.

5 https://asq.org/quality-resources/dmaic

6 FEMA has the Center for Domestic Preparedness (CDP), the Emergency Management Institute (EMI), and the National Training and Education Division (NTED). A course catalogue of NTED Courses may be found at: www.firstrespondertraining.gov/frts/npcc

7 See note 85 for Section 508 compliance.

Chapter 10

Developing the Risk Plan

A ship in harbor is safe, but that is not what ships are built for.

John A. Shedd, Salt from My Attic

RISK MANAGEMENT

We discussed risk management in Chapters 3 and 4 where we approached it from a community level (at various levels) and from the vantage point of an organization or government/government agency. Those chapters looked at risk in order to identify, evaluate, and prioritize potential projects to address risks. In this section on project risk management, we are looking at how risks can impact the project, a more focused viewpoint.

The diagram given in Figure 10.1 illustrates different levels at which risk can influence and ultimately impact a project. At the two outer layers, volatility, uncertainty, complexity, and ambiguity factor into the six aspects of working environment for organizations (collectively referred to here as STEEPLE):

- **Social** changes in communities, countries, and in global society as a whole
- **Technological** changes, including new systems, devices, and scientific discoveries that can have positive and negative impacts.
- **Economic** changes can influence an organization or project, buttressing or weakening supports.
- **Ecological**[1] the influence of and on the environment by the organization and project.
- **Political** changes, locally, regionally, nationally, and globally.
- **Legal** changes include regulatory and industry rules and guidelines.
- **Ethical** changes include how society may influence the organization and the organization's influence (including an industry) on society. This can include prominent community values (local/regional), morals, and behavioral norms.

These STEEPLE influences, affected by the general VUCA forces, are external to the organization and can impact the organization and in turn a project and the final outcome of the project, the product or service that is developed. We must also take into consideration how they interact between one another. We are going to focus on the two inner

DOI: 10.1201/9781003201557-12

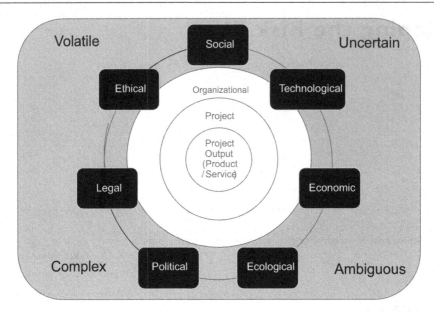

Figure 10.1 Project Risk Bullseye diagram.

circles of this framework, the project-based and "product-based" risk in this section. If you are reading this book out of sequence, then I would recommend going back to read Chapter 3, as well as Chapter 4 if you have time.

RISK MANAGEMENT PROCESS: OVERVIEW

Refer to Figure 10.2. While we covered the risk management process earlier, we are going to review this again a little differently when it comes to project risk. Note the following on the chart:

- The term "Risk Event" is used. As stated in Chapter 3, we can only a manage a risk event, once we have understood the factors that can cause it, triggers that alert us to the risk happening, and the impacts and consequences from the risk. To avoid any confusion, when I use risk and risk event below, they mean the same.
- The titles for each process are more descriptive of what takes place at each stage of the project risk management process. This is intentional, to help clarify what each process is designed to accomplish for the next step.
- It is a cycle. Risk management is an iterative process, even on a project until the project finishes, as exposures to risks and new risks may arise.

Again, let's walk through the process in a little greater detail. Please also note that terms and concepts covered previously will not be repeated at length here.

Figure 10.2 Risk management cycle.

IDENTIFY RISK EVENTS

We start off the risk management process by looking for sources of risk events and identifying specific risks. Where do we look?

- STEEPLE environment as presented above to understand the external risks to the organization and the project.
- Organizational risk scan (sometimes referred to as enterprise risk management[2]) which looks at: strategic risks, financial risks, operational risks, and hazard risks (including threats). This organizational historical record includes previous projects.
- Project plan elements: scope, schedule, budget, quality, procurement, human resources, stakeholder and communications. This often employs the use of a risk breakdown structure (RBS) similar to the WBS used in scope development.
- Stakeholders', including project team members, experience with risk.

The STEEPLE Analysis, an example of which is given in Table 10.1, can be helpful for working with senior leadership in the organization and with key stakeholders, especially with those who have a major interest in the project or have expressed concern over the viability of the project.

ORGANIZATIONAL RISK SCAN

The quadrants of risk is a model that originates from traditional risk management and can be particularly helpful for identifying risks on large-scale projects that are vulnerable to

Table 10.1 STEEPLE Analysis

Category	General Questions	Potential Risks
Social	What changes in social movements or cultural norms might affect the organization/project?	• A group of team members threaten to stop work over the need for greater equity hiring. • Communication issues between older and younger team members lead to incomplete definition of product requirements. • Community group initiates lawsuit for information on PII record storage.
Technological	What technological trends might affect the organization/project?	• Increase in global conflicts spurs greater cybersecurity attacks, leading to a cyberbreach during software launch resulting in software redevelopment. • State requirements for greater security and drive changes in server configurations, requiring changes to software specifications.
Economic	What economic factors might affect the organization/project?	• Inflation contributes to rising wages which leads to team attrition. • Increase in energy costs drive higher cost for rented servers.
Ecological	How might ecological (environmental) factors affect the organization/project?	• Client requires use of energy efficient computer equipment/servers for project which requires an upgrade by the company. • Local wildfire displaces team and client members, leading to delays in software development and reviews.
Political	How might the political environment affect the organization/project?	• Agency budget shortfalls cause a reduction in the scope of work for the project. • City's new mayor increases pressure on the emergency management agency to speed development on public service projects.
Legal	What legal, regulator/ compliance changes might affect the organization/project?	• City agency audit results in finding that company hiring practices are not in compliance with equal opportunity employment practices. • Software is not in compliance with ADA Section 508 requirements and requires rework.
Ethical	How might ethical factors affect the organization/project?	• A software developer on the team uses programming code from a previous company instead of developing it on their own, leading to possible copyright infringement.

changes in the organization and business environment. The following is an example of the quadrants of risk model that breaks risk into categories by type:

- **Hazard** based which are external to the organization, such as extreme natural weather events, cyberattacks, or power outages.
- **Operational** risks are internal processes and systems that fail due to human error, by accident, or by criminal intent.
- **Financial** risk is defined by any changes in the organization's or project's financial situation due to market fluctuations, for example costs of equipment and materials, stock price, and borrowing costs.
- **Strategic** risks are major changes in the marketplace, industry, or economy.

These quadrants also break out between pure risks on the top and speculative risks on the bottom. Pure risks are associated with negative outcomes, associated with losses, so only have downside to them. Speculative risks have an opportunity for gain or loss, so both an upside and downside. See Table 10.2.

The quadrants of risk are a helpful method to work with risk professionals and other executives in the company to help identify project risks and, when you begin developing solutions for prioritized risks under risk treatment and risk plan development, it will help identify some existing solutions and risk owners.

Table 10.2 Quadrants of Risk Model

Hazard	Operational
Building fire Derecho Earthquake Epidemic flu/contagious disease Extreme heat Flooding Power failure Wildfire	Contractor disruptions Internal network disruption Internal power failure Internal data theft Prolonged staff absences Software failure Staff accident Workplace violence
Financial	**Strategic**
Cash liquidity Change in funding Change in labor rates Company's Financial Profits/ Losses Interest rate changes	Company's change in product mix Contract dispute Change in market demand Change in company's media coverage Change in competitive market

Pure Risks

Speculative Risks

IDENTIFYING RISKS IN PROJECT PLAN ELEMENTS

Project risks can also be surfaced by having the team review the project plan as a whole: scope, schedule, budget, quality, procurement, human resources, stakeholder, and communications plans. This often employs the use of an RBS, similar to the WBS used in scope development. It also may necessitate an iterative approach to do a first pass on risk to identify viable options during the design stage, and then to repeat the process again once the additional project elements such as the quality and procurement plans have been developed.

An RBS as given below is a tool that can be used to identify and record risks. It can be used in conjunction with a brainstorming method that can help to cast a wide net on potential risks. Refer to Table 10.3.

You may want to consider using an RBS with the team, and one brainstorming technique that I like to employ with the team, and occasionally with the client when appropriate, is called a "Silent Brainstorming Exercise". As opposed to a regular brainstorming session, which is conducted as an actively facilitated open discussion of risks, this is done silently as follows:

1. The process may be facilitated by the project manager or other team member, or the participants can follow the rules and set the times themselves.
2. Participants gather around a table (square, round but at the same table). Each participant begins with a clean sheet of paper and pen, writes their initials at top, and is posed the questions:
   - What are the potential risks on the project, using their previous experience, and what they understand of the project?
   - What can go wrong on the project? What opportunities (positive risks) may arise?

Table 10.3 Risk Breakdown Structure

Technical	Organizational	External	Project Management
• Requirements • Technology • Complexity • Quality • Performance	• Project Dependencies • Resources-Budget, Funding, and Availability • Prioritization	• Contractors • Vendors • Customer • Working Environment	• Estimating • Planning • Controlling • Communication
• Changes to requirements based on what client wants. • Bugs emerge from server application stack. • Application does not work on all mobile devices.	• Internal senior managers demand project input or approval. • Organization reassigns team members to other priority projects. • Other projects may benefit from the application's feature set, so the project is subject to those requirements being defined.	• New Emergency Mgt. Agency Commissioner changes priorities. • Contractor proves to be unreliable in meeting deadlines. • Contractor produces poor-quality deliverables. • Contractor unable or unwilling to continue on project for any reason (emergency, bankruptcy, etc.) • Change in client Point of Contact • Other city agencies demand project input or approval. • Emergency incident causes work delays.	• Budgeted cost estimates are out of line with actuals. • Client does not engage with product reviews on a timely basis. • Schedule estimate is unrealistic, resulting in delays. • Delay in approval for project to start. • Delays in payments. • Project delay exceeds time funding is available. • Communication on complete version is not confirmed.

3. Each participant is given 3 min (more time can be allowed, especially on the first round) and writes down all potential risks they can imagine. I encourage each person to write neatly, so others can read what they wrote. When time is called, they pass the sheet to the next person.
4. Each participant reviews the list and considers additional risks. Again, 3 min is given to write.
5. This process continues until the list returns to the person who originated it. Each participant reviews their own list and checks for any duplication.
6. The facilitator (or group) asks one person to go first and read through their list.
7. As they read their list, the other participants listen and check off risks on their own list that have been identified. In cases where risks may differ slightly, they should not be checked off.
8. Each participant takes their turn (by asking the person next to the first speaker), and recites the risks that have not been identified till that point. The other participants listen and check off their lists in turn.
9. This process continues until all risks have been listed, without major duplication.
10. The lists are compiled and written up, putting them into categories according to the RBS, combined with lists of other project risks, and used as the basis for the next step in the risk management process.

One of the key benefits of conducting this brainstorming exercise silently is that it helps to elicit a more open and wide-ranging set of risks, especially from those participants who

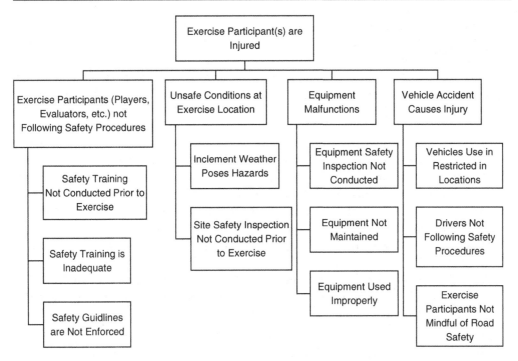

Figure 10.3 Fault tree diagram for a workplace injury.

may be more introverted. There are a number of methods and tools that can be used to identify and evaluate risks, and some of these are shared with quality management.

Similar to the Fishbone Diagram and the Bowtie Analysis, covered in Chapter 3 under risk management, is the Fault Tree Analysis, a tool to identify risks, as well as root causes. A Fault Tree starts with a major risk event and then breaks it down by potential factors which may contribute to triggering the risk event. Figure 10.3 is an example of a Fault Tree based on conducting a full-scale emergency exercise for a mass casualty event.

When it comes to analyzing a risk for treatment, which would be a further step in the risk management process following risk prioritization, each factor might be analyzed individually by looking at the probability and how it might contribute to a potential major risk.

Additional tools which have been covered previously and are often shared with quality management and strategic planning include:

- Value process model (similar to the SIPOC)
- Fishbone Diagram (Ishikawa; presented above in quality management)
- Assumptions and Constraints Analysis

Assumptions and Constraints Analysis looks at each assumption, derived from the scope statement and any brainstorming exercise, and each major constraint on the project and asking whether it could prove false, and, if false, whether it would impact the project. If both prove to be true, then it is converted to a risk.

Table 10.4 Assumptions and Constraints Analysis

Assumption or Constraint	Could This Prove False? (Y/N)	If False, Would It Affect Project? (Y/N)	Convert to a Risk
Sufficient, Qualified Exercise Evaluators	Y	Y	Y
Location is available and reserved for exercise use	N	NA	NA

A short sample is given in Table 10.4 of this method using a chart with only two factors based on the previous project example.

You may wonder why this evidently simple risk identification method may be needed, and in my experience often what is assumed can come back to haunt the project later. The following are some of the most common assumptions and constraints that may turn into risks on a project undertaken in an emergency management or public safety organization:

- All key stakeholders have been identified and are included in the review process. An example of this assumption leading to a risk might be legal review of any language included in planning, training, or other guidance documents, in particular qualifying what capabilities and services an organization is able to offer and limits to those services.
- Stable, well-defined product requirements. Quite frequently latent requirements are surfaced later in the development process that might have been identified earlier during the design stage, leading to costly rework. As stated in the previous section, quality methods can help to address this issue, as well as clearly outlining the requirements and development review process, including who is involved, number of review cycles, and deadlines. This also includes integration with existing systems.
- Contracts between the authorizing agency/organization and the implementing company/organization will be executed and funding made available on a timely basis. When this does not happen as planned and there are significant delays, team members may not be available or their availability may be limited, development time may be constrained, which can lead to lower quality production. External, third party, procurements, and contract approvals also need lead times which can cause risks to final product delivery.
- Future maintenance and support for the product, e.g. operational guides, web-based applications, e-learning courses, especially when it comes to future configuration or changes in regulatory requirements. These are issues which can be becomes risks during the transfer of final deliverables to the client.
- Response to large-scale emergency events will take time away from the project. This may seem obvious, but I have found that in this age of flat and lean staffed organizations, a major incident can pull client teams away from development meetings or major reviews, leading to significant delays.

Some additional areas to explore risks in greater detail are:

- Project schedule for multiple predecessor or successor dependencies, dependencies on approvals.

- Scope of work, including the number and complexity of requirements.
- The budgeting cycle and funding flows and expected delays.
- Stakeholders – numbers and level of influence and experience or history with them.
- Quality requirements, including previous project reports, end of project punch lists, and end-user requests.
- Lessons learned, including loss reports, delays, and client reviews, from previous projects.
- Vendor/contractor/supplier reviews and any references.

You may have noticed that using multiple methods in the risk identification process can identify risks multiple times, an apparent redundancy of effort. This is a natural part of the process and ensures comprehensive coverage and gaining a greater vantage point on risks, including their root factors and potential impacts. All risks should be considered at this stage of the process and should not be left off the list until they are assessed for prioritization.

ASSESS RISKS FOR PRIORITIZATION

So what is the key difference between risk assessment and risk analysis? At this point in the risk management process, you would have accumulated a long list of risks in your RBS or other documentation, and the next step is to start assessing which of these risks are critical to address and which are less likely to have a major impact on the project. Risk assessment at this stage of the process is intended to be a basic process, designed to differentiate which risks are more important to consider for analysis and which require less attention. It is not an in-depth analysis as used in the next step of the process intended to thoroughly understand the full impacts and consequences of risks and identify potential ways to treat those risks. Basically, we do not want to spend time conducting a rigorous analysis on risks which are of little consequence to the project.

Our first step is to assess each risk by their probability and impact, and the most common method is using a Risk Rating Matrix. An example is given in Table 10.5 using the same example from above.

There are a couple of features of the Risk Rating Matrix above that are important to point out:

- It uses a three-point scale (High=3, Medium= 2, Low=1) for both probability and impact. Some rating matrices may use a finer scale, 1–5 or 1–10 or have different scales for probability and impact. Some may use a percentage for probability and another for impact (see Chapter 3 for risk assessment methods).
- Both of the scales are used to create a final rating by simply multiplying the two by one another.
- The risks are not analyzed by the level of vulnerability of the project or organization, nor are they treated (addressed by mitigation measures); that will come in the next steps.

Table 10.5 Risk Rating Matrix

Risk No.	Risk Description	Probability (Low-Med-High)	Impact/Cost (Low-Med-High)	Rating
1	Delay in approval for project to start	3	3	9
2	Contractor unable or unwilling to continue on project for any reason (emergency, bankruptcy, etc.)	3	3	9
3	Contractor produces poor-quality deliverables	2	3	6
4	Emergency incident causes work delays	2	3	6
5	Changes to requirements based on what client wants	2	3	6
6	Contractor proves to be unreliable in meeting deadlines	2	3	6
7	Other city agencies demand project input or approval	2	3	6
8	Client does not engage with product reviews on a timely basis	2	3	6
9	Project delay exceeds time funding is available	2	3	6
10	Organization reassigns team members to other priority projects	2	3	6
11	Other projects may benefit from the application's feature set, so the project is subject to those requirements being defined	2	2	4
12	Budgeted cost estimates are out of line with actuals	2	2	4
13	Change in client Point of Contact	1	3	3
14	New Emergency Mgt. Agency Commissioner changes priorities	1	3	3
15	Schedule estimate is unrealistic, resulting in delays	1	2	2
16	Delays in payments	2	1	2
17	Bugs emerge from server application stack	2	1	2
18	Application does not work on all mobile devices	2	1	2
19	Internal senior managers demand project input or approval	1	2	2
20	Communication on complete version is not confirmed	2	1	2

We obtain the information on probability and impact from some of the same sources as we would quality information from historical project experience:

- Lessons learned (AAR/IP) reports from previous projects
- Interviews with subject matter experts (SMEs), experienced project managers, vendors, and subcontractors
- Discussion with team members
- Data from recent hazard or emergency events
- Complaints from clients/customers (may also be end-users)
- Contract issues from past projects/services
- Product logs, e.g. software bug tickets and reports

As with other aspects of project planning, the project team should be closely involved in the risk assessment and establishing a consensus on the priorities. Referencing the prioritized list of risks above, if the organization has experienced delays in the approval for the project start (risk#1), and it has had a major impact on the project (think major delays and potential increase in level of efforts and costs), then it stands to reason that this would be prioritized over the risk of a delay in payment (risk#16). This may be based on the knowledge that this is less frequent a risk and staff will still be paid due to sufficient cash on hand despite any delays.

As an example of the end-product of the risk assessment process, the team may deem that risks number 15–20 warrant less attention than the ones above that rating threshold and will be addressed through other elements of project planning, such as quality, human resources, procurement, and plans. These may still be monitored during the project execution but perhaps not analyzed and treated in-depth as with the prioritized risks.

Back-to-Back Hurricanes in New York City

Hurricane Irene made landfall as a tropical storm with its eye travelling directly over Coney Island in Brooklyn on August 29, 2011. While it wrought tremendous damage in upstate New York and Vermont, it did much less damage in the New York/New Jersey metro area. Nevertheless, full preparations were made for a category one hurricane starting four days prior, with evacuations from coastal areas and emergency sheltering in full swing during the 48–24h period before tropical storm winds started whipping the tristate area. Following Hurricane Irene, numerous After-Action Review meetings took place among various City agencies and the Mayor's office to consolidate lessons learned and to make future improvements. One key area of improvement among these was in the evacuation and emergency sheltering functions.

After wreaking havoc in the Caribbean in late October of 2012, Hurricane Sandy emerged from the Turks and Caicos heading north, while most hurricanes blow out to sea in the Atlantic and become a "fish storm", Sandy made an unusual left turn toward the New York/New Jersey coast, transformed into a mix of a tropical storm and Nor'easter creating a Superstorm, the equivalent of a category one, when it made its final landfall just north of Atlantic City, New Jersey on October 29th. This time the New York City area felt the full brunt of a true hurricane, an event which had not happened since 1938.[3]

The probability of two hurricanes taking place in succession one after the other, in the same area is about one in 2,500 in any two-year period.[4] The emergency preparations for Hurricane Irene provided a dress rehearsal for Hurricane Sandy and allowed many plans to be practiced and systems tested in an actual response. This proved to be a major opportunity, a positive risk, to enhance the capabilities of the City's emergency response infrastructure for the next major event. It also presented a negative risk, as the response to Hurricane Irene in 2011 delayed by months many development projects, such as training, software application development, and facility upgrades.

One specific lesson learned was in application delivery of critical software used in emergency response. Cloud-based servers, which were relatively new at the time, provided a greater level of resilience against system outages during a major hurricane than local servers, which were primary, with cloud-based servers used as a backup. By the time Hurricane Sandy arrived in late 2012, cloud-based servers became primary for some of the critical software applications used with emergency response, with distributed local servers used as a backup. This new configuration proved a resilient solution, delivering 99.9% uptime despite the 1,000-mile-wide storm field and extensive power and system outages from Maryland to Massachusetts and as far inland as West Virginia.

ANALYZE RISKS FOR TREATMENT

Similar to how risk analysis was addressed in Chapter 3 on Risk Management for Emergency Management and Public Safety Planning, we are taking the prioritized risks and performing a more in-depth, quantitative analysis, except this time, in order to better understand the vulnerabilities and impacts from risks to the project we are planning so that we can treat the risks to reduce the probability and impacts they may have on the project.

Before we move forward with Risk Analysis, I would like to introduce one of the most common tools used by both Project Managers, as well as Risk Managers, and that is the Risk Register. While there are many different formats for a risk register, I have found the following categories to define a risk the most useful, and they are identified and explained further below. Again, we will use an example that follows on from the software development project. Refer to the example in Table 10.6. Risk Register Example: Software Application for Client Record Registration for Special Medical Needs Shelter

The risk register items given in Tables 10.6 and 10.7 for analysis may not be the only areas we may want to explore. As we covered earlier in Chapter 3 on risk, some additional areas of risk might include:

- **Time horizon:** duration that the risk may last. This is especially important when it comes to emergency incidents, a major storm may last for days and recovery for days to weeks, depending on its size, vs. a limited gas leak that may impact the project for hours to days during the response.
- **Correlation:** relationship between risks or risk factors. Risks do not happen in a vacuum and may influence one another. As we have in this example, if the client is an emergency management organization and there is a prolonged emergency incident, their product reviews may be delayed and the POC may be unavailable and reassigned to someone else in the organization temporarily.
- **Geographic range:** how geographically widespread the risk is. A good example of this is an epidemic or pandemic which can have a systemic impact to a wider group of partners, suppliers, and on the client team and much deeper impacts to a project.
- **Volatility:** to what degree is the risk event subject to change once it has occurred, becoming amplified, waning, or transforms into another type of unexpected event. A good example of this is Hurricane Harvey, which was a category 4 hurricane and thought to be typical in behavior that would pass through in 12–24 h and become a high wind event with flooding effects from storm surge. Instead, Hurricane Harvey lingered for four days dropping 60 in. of rain and making it one of the wettest, bringing flooding farther inland, making it one of the costliest hurricanes in US history.[5]
- **Vulnerability and exposure to the risk:** to what degree does the organization have a vulnerability to the risk, based on its facilities, systems, and staff. It also requires us to look at what is already in place, insurance coverages, security measures, and other protective measures that may address project risk exposures. This would likely include an organized review of the project with professional risk managers in the organization.
- **Risk treatment currently:** this would include any risk transference through insurance coverage or contracts, internal risk controls, or other existing organizational measures which reduce risks.
- **Residual risk:** this is the amount of risk after risks have been addressed through risk measures. Initially this will be through current measures and ultimately by all planned measures.

Table 10.6 Risk Register for Special Medical Needs Shelter Client Record Registration-Software Application

Risk No.	Risk Description	Risk Trigger and Timing	Owner	Probability	Impact Description	Cost Impact	Schedule Impact	Response Description	Response Type
1	Delay in approval for project to start	Contract approval delayed, Finance/admin. failure to deposit funds in account; longer notice[a]	Project Manager	75%	Reduces the time for work on the project; may impact team members available.	May increase costs; estimated increase in OT[b] = 15%	Reduces time allowed for work, by days/weeks.	Work with legal counsel and finance to address obstacles. Increase project schedule by anticipated delay.	Mitigation, Proactive
2	Contractor unable or unwilling to continue on project for any reason (emergency, bankruptcy, etc.)	Contractor fails to deliver service; official notice; comments from contract staff member, limited notice	Procurement Director	60%	Need to find an alternative service provider on short notice, with increased costs	May increase costs for a replacement by 10%	Create a delay in development by days/weeks	Identify backup contractor; include penalty clause in payments to account for added costs	Mitigation, Proactive; Risk Transfer, Reactive
3	Contractor produces poor-quality deliverables (e-learning courses)	Failed production test (fails in LMS), limited to longer notice	Quality Manager	50%	Poorly developed e-learning courses require major rework	Penalty of 20% in delivering final software version	Delay in software rollout due to software training support	Conduct due diligence on vendor selection; include penalties for non-performance	Mitigation, Proactive; (action 2) Risk Transfer, Reactive
4	Emergency incident causes work delays	Alert of a natural hazard – Notice (storm/hurricane); No or limited notice (tornado/earthquake/wild-fire) Alert human threat/accident – No notice (power outage, active shooting event)	Project Manager	50%	Client reviews and testing are delayed by weeks/months; may impact scope	May increase costs by up 15–20% due to need to for OT or using new team members; if scope is impacted, may increase costs	Delay ranges from weeks to months	Add contingency time based on past emergency incident response history; Work with legal counsel and finance to extend the deadline (no-cost); if scope impacted, then renegotiate scope of work	Mitigation, Proactive; Response, Reactive
5	Changes to requirements based on what client wants	Communication of change request by client (review feedback, email) limited notice	Lead Software Developer	50%	Change in product requirements increases scope of work	Dependent on change, historic data shows increase in costs = 5%	Delay ranges from days to weeks	Build early prototyping into development and review cycle; If scope impacted, then renegotiate scope of work	Mitigation, Proactive; Response, Reactive

(Continued)

Table 10.6 (Cont.)

Risk No.	Risk Description	Risk Trigger and Timing	Owner	Probability	Impact Description	Cost Impact	Schedule Impact	Response Description	Response Type
6	Contractor proves to be unreliable in meeting deadlines	Misses more than one deadline, limited notice	Procurement Director	45%	Causes delay in final product delivery	Low-cost impact	Major delay in development and acceptance of final deliverables, by weeks	Due diligence during the vendor selection process; contract penalties for delays in delivery; identify alternative vendors.	Mitigation, Proactive; Risk Transfer, Reactive
7	Other city agencies demand project input or approval	Communication by agency, limited notice	Project Manager	45%	Causes reevaluation of project scope and application requirements	Dependent on change, historic data shows increase in costs = 5%	Delay ranges from days to weeks	During early stakeholder management address showstoppers; negotiate with external agency to accommodate demand	Mitigation, Proactive; Response, Reactive
8	Client does not engage with product reviews on a timely basis	Failure to participate at review meetings or provide substantial feedback for reviews, limited to longer notice	Lead Software Developer	40%	Requirements not full defined and approved	Dependent on change, historic data shows increase in costs >= 5%	Delay ranges from days to weeks	Engage with client team early on and require sign-off at key development stages	Mitigation, Proactive
9	Project delay exceeds time funding is available	Halfway through the schedule estimated time to complete project exceeds the funding deadline, longer notice	Project Manager	40%	Funding not available to complete remaining work	Increase in cost for additional resources or OT; Revenue loss (>-5%)	Less time available to complete the remaining work	Build in buffer time (5% of schedule) for delays; request no-cost extension.	Mitigation, Proactive; Response, Reactive
10	Organization reassigns team members to other priority projects.	Communication from team member, HR, or department head, no notice or limited notice	Project Manager	35%	Delay in work while a replacement is found (dependent on role)	Hiring and training cost of replacement (may be indirect to organization)	Delay on task completion and project schedule	Cross train team members; develop succession plan in company; negotiate with department heads for delay in reassignment	Mitigation, Proactive (first two); Response, Reactive

#	Risk	Source	Owner	Probability	Impact	Threshold	Trigger	Response	Type
11	Other, internal projects may benefit from the application's feature set, so the project is subject to those requirements being defined	Communication from department heads, chiefly the Project Management Director, longer notice	Lead Software Developer	35%	Significant delay and additional work in integrating other requirements into the project	Increase in development costs (< 5% increase); if opportunity may increase revenue on other projects	Delay on project schedule of >5%	Negotiate to either: continue with project as scheduled; or negotiation with client to allow more time and share opportunity	Mitigation, Proactive; Share the opportunity (positive risk) with internal depts., reactive
12	Budgeted cost estimates are out of line with actuals	Contracted services exceed estimates; actual costs exceed budgeted costs on project financial reports, longer notice	Project Manager	30%	Projected project cost exceeds budget	Costs exceed 5% of budget; reduce profit or cause a loss on the project	NA	Obtain quotes for contracted service when developing the budget; allocate a contingency budget of 5% for cost overruns	Mitigation, Proactive
13	Change in client Point of Contact (POC) and review of project scope and requirements	Communication from client, longer notice	Project Manager	25%	Possible review of project scope and product requirements; change in support of project (may be positive or negative); depends on timing with schedule	Changes in scope would have cost implications	Delays in days/weeks; possible cancelation	If change in project scope and requirements, negotiate with new POC	Mitigation, reactive
14	New Emergency Mgt. Agency Commissioner changes organization priorities	Communication from client, longer notice	Project Manager	25%	Possible review of project scope and product requirements; change in support of project (may be positive or negative); depends on timing with schedule	Changes in scope would have cost implications	Delays in weeks/months; possible cancelation	If change in project scope and requirements, negotiate with new POC/new commissioner	Mitigation, reactive

a   No notice means no lead time; limited notice may be hours, days, or weeks (duration is approximately six months for this type of project); longer term notice may be weeks or month in advance.

b   OT = overtime.

c   Showstopperᶜ = a situation that cause the project to come to an abrupt stop, usually due to a critical requirement that is not being met and being called out by someone who has the authority to call for the stop.

We will now describe through each of the items in the Risk Register given above using Table 10.7.

Table 10.7 Risk Register Field Title Definitions

Risk Register Item	Description
Risk No.	This number can be based on the initial prioritization matrix or use a number from the RBS (similar format to a WBS). This is used for tracking risks more easily.
Risk Description	This is a short description of the risk. As stated earlier, this needs to be stated as a specific risk event as clearly possible.
Risk Trigger and Timing	This is the most clearly defined trigger that indicates that the risk event has taken place. It also includes the approximate time in advance that a warning the risk may take place. This equates to the speed of onset, or velocity, that we discussed briefly in Chapter 3. Note that the time that a warning might be received may vary in ranges, and for some it may be challenging to specify this. The example given above identifies three categories of timing: no lead time; limited notice, and longer notice, and time ranges are provided for these terms. For your own purposes, you may decide to define time frames more or less precisely depending on the needs of the project.
Owner	This is the position on the project team that is responsible for planning for and managing the risk. In cases where a team member cannot assume responsibility, the Project Manager assumes responsibility. While some risks may be escalated to the internal Project Sponsor (in this example it would be the Project Management Director), the direct responsibility belongs to the project team.
Probability	This is the probability that the risk will take place during the project. In this example, we are using the percentage probability from 0 to 100%, and this would be derived from research on past projects.
Impact Description	This is a description of what impact the risk would have on the project. It may include additional impacts such as time and cost, although additional categories are provided for these (see below).
Cost Impact	This is the estimated cost implications in percentage of the budget or in true monetary terms. When it is possible, it is best to include the estimated cost if this risk occurs. In some cases, this may be a range, for example theft of physical items or a cybersecurity event may have a potential range in cost impact.
Schedule Impact	Similar to cost impact, this is the estimated impact to the schedule and may be stated in a range of time, in this example using days, weeks, to months.
Response Description	This is a description of the action or actions that may address the risk. We will discuss this part of the risk register in the next sections.
Response Type	This is a description of the type of risk response, whether this is proactive or reactive; mitigation, transfer, acceptance, or avoidance. Most of these risks are addressed with either mitigation or transference, as opposed to merely accepting the risk. We will describe response type in the next sections.

Some impacts from risks may also be taken into consideration that are difficult to measure objectively, such as changes to organizational reputation, team morale, and client expectations. However, it may be useful to include these more qualitative impacts by creating some notional measures around them. Options from a well-designed survey instrument to a simple poll among respondents might provide a method to measure these qualitative risk impacts among stakeholder groups.

How do we collect data to inform risk measurements such as probability and impacts to schedule, cost, and scope? Most often, the project team will conduct initial risk research based on recent projects, striving to be as objective in their analysis.

Some additional avenues to explore risks:

- Interviews and surveys with specialists (SMEs); for example, if we are looking at the risk of cybersecurity to our project, then we might consult with the Chief Information Security Officer (CISO) for how often these events rise to a risk event.
- Looking at product claims, insurance losses, number of sick days, and actual costs from recent events, again turning to professional risk managers for data and analysis.

BOWTIE ANALYSIS REVISITED

We are going to revisit the Bowtie Analysis, as this is an effective useful tool for risks that are more likely to take place and have a greater impact. It provides a method that is easy to use and to dissect a risk event and also helps feed the next step in the risk management process: treat risks and develop the risk management plan. We will analyze one risk from the previous analysis: delay in approval for project start. Refer to Table 10.8.

As we discussed before, the causal factors are a preceding event or condition that contributes to the occurrence of a risk event. The direct consequences are the impacts that the risk event can create. We can break down causal factors to their own root causes as illustrated in the Fault Tree Analysis under risk identification. In a similar fashion, we can continue to break down the consequences of a risk event into an Event Tree Analysis, identifying cascading events into first-order, second-order, and third-order impacts, and so forth. An example is presented in Figure 10.4 for a Riverine Flooding Event.

Table 10.8 Bowtie Analysis

Causal Factor(s)	Mitigation Measures	Event	Risk Response Measures	Direct Consequences
• Delay in legal contract execution between parties • Delay in approved funding allocated to account • Scope of work not confirmed by client POC • Workplan not approved by client POC • Kick-off meeting delayed by client team	• Add buffer time in schedule for contract and funding allocation delays • Track contract with legal to ensure contract is executed and logged • Track financing and coordinate with finance staff to ensure funding is available on time • Manage communication and coordination with client POC and team to ensure workplan is approved and kick-off meeting is scheduled on a timely basis	Delay in approval for project start	• Request no-cost extension to meet original schedule • Add additional team members to critical path tasks to shrink the schedule • Renegotiate the scope of work to reduce the time needed to complete the project	Less time to complete the scope of work which may lead to: • Push to complete tasks in a shorter amount of time may create quality defects in the product • Team burnout working longer hours may cause a loss of team members or underperformance • Inability to deliver all of the product functionality according to the defined requirements • Increase in costs due to additional staff resources to complete tasks on time • Funding not available for completing the project work

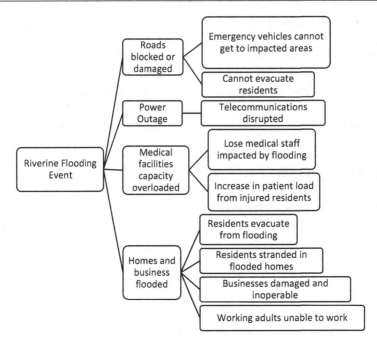

Figure 10.4 Event tree diagram. This is a simple example and does not represent all possible outcomes including primary and tertiary events.

EVENT TREE DIAGRAM FOR A RIVER FLOODING EVENT

The primary event is the River Flooding which cause major impacts: roads blocked or damaged, medical facilities overloaded, homes, and businesses flooded (first-order effects), and the second-order effects that emanate from them. This is just a simple example, and we could imagine identifying further effects from the second-order effects, for example the consequences of residents who are stranded in their homes due to the flooding or businesses which are inoperable from the flooding. Again, we are looking at this from the point of a project, so if we were thinking about the risk above in terms of the software development project, then we would look at Liberty City and how this might impact our project: the client agency, our team members and organization, including the offices and the surrounding infrastructure.

MONTE-CARLO ANALYSIS, MODELING, AND SIMULATION

For larger projects with greater value at risk (multiple of millions of dollars, euros, pounds sterling, etc.) and longer duration such as large-scale engineering or software projects, more sophisticated methods may be employed such as modeling software or statistical simulation, including Monte Carlo analysis. Monte Carlo analysis is an analytical method used for complex investments (typically in the area of finance), projects, or systems that involve many areas of uncertainty. Project inputs can be quantified with a specific range then run thousands or tens of thousands of outcomes to yield what is the range of most likely outcomes.

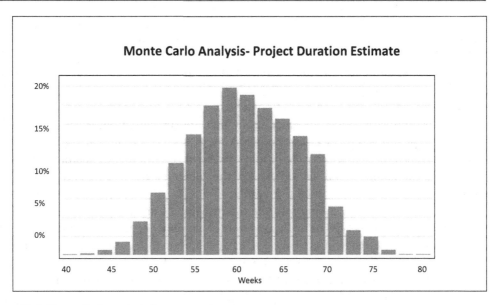

Figure 10.5 Monte Carlo analysis for project duration.

An example is provided in Figure 10.5 for a schedule based on probabilistic estimating (three-point estimates using optimistic [O], realistic [R], and pessimistic [P] estimates for each task[6]; formula would be [(O+4R+P)/6] for a weighted average toward the middle), running three-point estimates for each task multiple times and then totaling the project duration for each round.

The histogram above represents running a Monte Carlo analysis thousands of times for a project schedule in weeks, each number of permutations that fall within a schedule duration range is represented by a bar and the X-axis (along the bottom) identifies the approximate number of weeks for each. The Y-axis (on the left side) shows the percentage of the total number of iterations run. We can see from the analysis of potential project durations that the most optimistic schedule is around 40 weeks, but less than 1% of the outcomes fall within this range. On the other end of the scale, we have the pessimistic estimate of 80 weeks, again with fewer than 1% of the outcomes falling within this range. The average of all estimated project durations is around 60 weeks.

How does this type of analysis come in handy? Depending on the risk strategy of an organization, the range of schedule estimates and frequency might be used for creating the planned schedule as part of a proposal. If the organization wants to be risk averse and ensure that they are reasonably confident in the project finishing on time, they may choose the most pessimistic estimate of 80 weeks for the project given above. If on the other hand an organization wants to be risk taking (sometimes referred to as risk accepting or risk prone; these are aggressive risk approaches), then they would choose a project duration closer to 40 weeks. The same method of analysis can be done for cost estimates as well as any aspects of the project that may have a range of outcomes, such as risk events or quality outcomes.

Modeling may involve using software or systems that will model project plans or designs to better understand how they might perform. For example, it is common in the insurance, construction, and engineering industries to use hazard models in areas where natural hazards such as hurricanes, tornados, or earthquakes may be encountered. Modeling that integrates some of the common commercial tools, such as Building Information Modeling (BIM) and geographic information system (GIS) mapping data, are now being used to create greater resilience in the design phase.[7]

The use of simulations is familiar to emergency and public safety professionals, who may use this in their training and exercises in high pressure situations, such as firefighter training on apparatus, EMS workers on manikins, or law enforcement in critical incident simulators. For projects that may entail some degree of risk, high-cost, and/or complexity, conducting simulations might help to analyze the vulnerabilities and potential outcomes, especially the interactions between risk events.

A note of caution when it comes to the use of Monte Carlo Analysis, simulations, or modeling methods. This type of analysis is only as good as the quality of the data going into it (the saying "garbage in, garbage out" is appropriate thought to keep in mind). The quality of optimistic and pessimistic estimates rests on the experience of the team and organization, with many project-based factors to take into consideration. In addition, organizational culture and attitude toward risk play a significant role in the decision-making process when using these methods. A senior manager in an organization may have a very optimistic outlook and ignore the pessimistic ranges or potentially negative outcomes, assuming risks without understanding the full ramifications.

EXPECTED MONETARY VALUE AND DECISION TREES

In the risk assessment phase, we rated risks by their probability and impact using a qualitative scale. In the analysis phase, we can use a similar method, a quantitative one, to arrive at what is the expected monetary value (EMV) based on probability and a cost impact. EMV is a risk-based decision-making method that uses statistics (probabilities) and estimated cost outcomes for projects and options to pursue in projects. The basic formula for EMV is:

Probability (0–100%) × Cost Impact (in \$, €, £, etc.) = Expected Monetary Value

As an example of this method, let's use risk 10 from project risk register above "Organization reassigns team members to other priority projects" with a probability of 35%. The cost impact is "Hiring and training cost of replacement". For our purposes, we will use an estimate of five months average salary to replace and train a team member, with \$120,000 as the estimated compensation (salary+ fringe costs) per year= \$50,000. We have as our EMV

Probability 35% × Cost Impact \$50,000 = Expected Monetary Value of \$17,500

What does this method tell us about this and the other identified risks on the project? When we are planning for all risks, we need to consider how we "pay down these risks", in other words how we decide to treat them in order to reduce the potential loss. We can do this for all of the risks we included in our risk register and, after risk measures that are already in place, we can look at the residual risk, in other words the untreated risks for what their EMVs are individually and collectively. This also provides an opportunity for the project team to resort the order of priority once a rigorous analysis has been completed.

An example using the top five risks from the risk register is given in Table 10.9 on how this can be used for project planning. For purposes of our sample project, we will work with an estimated project budget of $1 million (again, this is a fictional amount) and focus solely on costs and not on schedule impacts. Note – we have removed some of the category columns from the risk register to simplify the EMV calculations and added columns for EMV, current risk treatments, and residual risk. For some risks, there will be no current risk treatments, and for estimated cost impact ranges, we use the greater amount.

In our simplified example using the top five prioritized risks, we see that we have approximately $400,000 in estimated monetary value of untreated risks, with a little more than $200,000 in risk treatment, with a residual value of $192,500 EMV left over. We can also see that risk 3 has a greater EMV value than risk 2, it after so can be raised in priority. We can perform the same analysis with the remaining risks in our risk register to obtain a more complete analysis of the total risks. EMV provides a quantifiable basis for identifying the range that we may wish to pay down risks as we alluded to before. Clearly, we would need to treat the residual risk amount, which we will address in the next section.

The last method that we will address in risk analysis is the use of a Decision Tree based on probabilities and costs using the method of EMV. The calculations are the same, and the method allows us to use quantitative analysis to objectively choose between multiple options when there is a limited number of alternatives. An example is provided in Table 10.6 for a decision on how to address a staffing shortage within the team. The two alternatives are for hiring a UX Specialist internally or using a contract staff person from a placement agency.

Refer to Figure 10.6, which presents a simple model of two decision options with two possible outcomes: that the new hire works out well or does not work well and that we encounter additional costs to hire a replacement. We calculate each outcome and then multiply the probability by the total costs for that outcome. We then add each of the outcomes for that branch of the decision tree for the total EMV. Evaluating whether it make more sense to hire an internal UX specialist or to contract for one through a staffing company, we can come to the conclusion that it is less costly to use a contracted staff person.

However, the figure above uses a full year's salary for the internal staff hire because the company requires that for project-based accounting, even though this position may be needed for nine months on the project. As this is a clear business decision, this might be a discussion with the Project Management Director to see if the other three months of the time might be borne by other projects or if there is a cost-sharing with other departmental work that may be undertaken. Furthermore, we are only looking at two options here. Some staffing companies offer an option to hire a contract staff at an added cost. These factors may be taken into account for this decision.

As we laid out before, in constructing a decision tree like the one Figure 10.6, we would want to ensure that we are using reasonably accurate historic data, so we would consult with our human resource department (and possibly the procurement department,

Table 10.9 Expected Monetary Value Analysis

Risk No.	Risk Description	Probability	Impact Description	Cost Impact	EMV	Current Risk Treatment	Residual Risk
1	Delay in approval for project to start	75%	Reduces the time for work on the project; may impact team members available	May increase costs; estimated increase in OT** = 15%	75% ×15% × $1 mill. Budget = $112,500	Risk is not currently treated	$112,500
2	Contractor unable or unwilling to continue on project for any reason (emergency, bankruptcy, etc.)	60%	Need to find an alternative service provider on short notice with increased costs	May increase costs for a replacement by 10%; note: subcontract value is $200,000	60% × 10% × $1 mill. Budget = $60,000	Current subcontract terms include a penalty of 5% for every two-week delivery delay and penalty for canceling the contract for cause at 20% of total = $40,000	$20,000
3	Contractor produces poor-quality deliverables (e-learning courses)	50%	Poorly developed e-learning courses require major rework	Penalty in delivering final software version= 20% of total project value; note: subcontract value is $200,000	50% × 20% × $1 mill. Budget = $100,000;	Penalty for canceling the contract for cause at 20% of total= $40,000	$60,000
4	Emergency incident causes work delays	50%	Client reviews and testing are delayed by weeks/ months; may impact scope	May increase costs by up 15–20% due to need for OT or using new team members; if scope is impacted, it may increase costs	50% × 20% × $1 mill. Budget = $100,000	Include contract clause for a no cost extension in case of delays due to major emergency incidents (details of which spelled out in the contract). Organization has business interruption insurance	No cost for an extension, but scope changes remain a residual risk (cost unknown)
5	Changes to requirements based on what client wants	50%	Change in product requirements increases scope of work	Dependent on change, historic data shows increase in costs = 5%	50% × 5% × $1 mill. Budget = $25,000	Include contract clause for any major changes in scope (details of which are spelled out in the contract)	Residual risk of scope changes still needs to be managed
				Total	$397,500	Approx. $205,000	Approx. $192,500; residual risks for scope changes

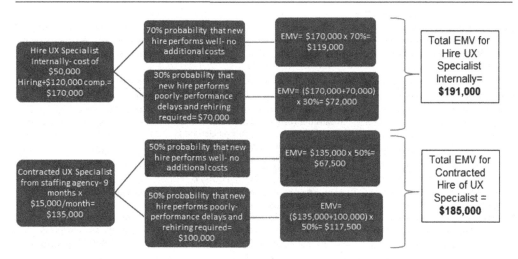

Figure 10.6 Decision tree analysis.

depending on who hires contract staff) to obtain accurate information on the performance and attrition levels for internal vs. contract hires.

Referring back to Chapter 3 on Risks, we addressed the Cost Benefit Analysis of which Decision Tree analysis is a form. These methods can be used for scenario analysis, looking at potential cost inputs and benefit outcomes and their probabilities for each potential scenario.

TREAT RISKS AND DEVELOP RISK MGT. PLAN

Now we get to the key reason we undertake risk management: to treat the risks through measures that will address the residual risks that the project may face and develop the risk management plan, with the ultimate objective of achieving the project goal and deliverables as planned.

We are going to revisit risk response strategies, except this time we will cover major strategies for both negative and positive (opportunities) project risks, refer to Table 10.10.

Note that except for acceptance and avoidance strategies, these risk measures are not mutually exclusive and may be taken collectively if the mutual strategies can be integrated and are advantageous to the project.

Response can also be grouped into phases before and after the risk occurs, which can be looked at as proactive or reactive depending on which side of the risk event. Refer to Figure 10.7; from this type of analysis, we want to invest as much as is reasonable from a cost/benefit perspective to eliminate the probability of a risk occurring or reducing its impact once it has occurred through mitigation measures.

With both frameworks in mind, let's revisit the risk register with the top five risks, adding risk No. 11, since this is the only opportunity (positive risk) for the organization that we have identified so far. The planned responses are presented in Table 10.11.

Table 10.10 Risk Response Strategies

Applicable to Negative Risks or Positive Risks	Risk Strategy	Explanation	Example
Both negative and positive risks	Acceptance	No action is taken to avoid or address the impact from a risk if it happens. This means the response will be taken at the time it happens. No contingencies are planned in advance, requiring added time or cost to the project	Not planning for a loss of team members
	Escalation	Escalate a risk event to a more senior management level when authority for responding exceeds that of the project manager or project team	The client presents an opportunity to pay for added product features
Negative risks	Avoidance	Eliminate the risk by taking a different course of action or removing a part of the project scope that does not incur the risk	Not including a long-term maintenance contract to support a software program
	Prevention/ mitigation	Reduce the probability or impact of the risk	Creating a backup server site for software production
	Transference	Transfer responsibility for the risk to a third party through purchasing insurance or outsourcing to another organization	Purchasing cyber insurance for an IT project
Positive risks	Exploiting	Pursue the opportunity by taking action to enhance the value for the project	Creating ADA-compliant facilities for emergency use provides benefits for families with children and for everyday use
	Enhancing	Optimizing the factors that influence the opportunity to ensure the probability of a positive outcome for the project	Reviewing the schedule with the vendors and client to find ways to reduce the project schedule
	Sharing	Collaborating or cooperating with other projects, departments, vendors, or client to exchange ideas or leverage resources to share in the opportunity	Negotiating with vendors used by other projects to use their services to reduce contract costs

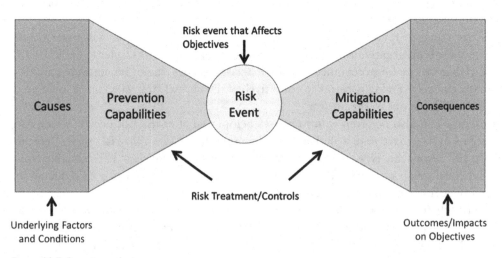

Figure 10.7 Bowtie analysis.

Table 10.11 Risk Register-EMV and Risk Treatment Analysis

Risk No.	Risk Description	Impact Description	EMV	Risk Treatment	Type of Risk Response	Residual Risk
1	Delay in approval for project to start	Reduces the time for work on the project; may impact team members available.	75% × 15% × $1 mill. Budget = $112,500	• Increase project schedule by anticipated delay. • Work with legal counsel and finance to address obstacles.	Prevention and mitigation. Proactive	Reduced the probability to a third= $37,500
2	Contractor unable or unwilling to continue on project for any reason (emergency, bankruptcy, etc.)	Need to find an alternative service provider on short notice, with increased costs	60% × 10% × $1 mill. Budget = $60,000	• Current subcontract terms include a penalty of 5% for every two-weeks delivery delay, and penalty for canceling the contract for cause at 20% of total=$40,000 • Request that the vendor penalty for poor performance amount to $60K	Mitigation/ Proactive	0
3	Contractor produces poor-quality deliverables (e-learning courses)	Poorly developed e-learning courses require major rework	50% × 20% × $1 mill. Budget = $100,000;			0
4	Emergency incident causes work delays	Client reviews and testing are delayed by weeks/months; may impact scope	50% × 20% × $1 mill. Budget = $100,000	• Include contract clause for a no cost extension in case of delays due to major emergency incidents (details of which spelled out in the contract). • Organization has business interruption insurance. • Scope changes are negotiated.	Transference/ Prevention (1, 2); Mitigation/ Response (3)	0
5	Changes to requirements based on what client wants	Change in product requirements increases scope of work	50% × 5% × $1 mill. Budget = $25,000	• Include contract clause for any major changes in scope (details of which are spelled out in the contract). • Conduct rapid prototyping to elicit any latent requirements	Prevention/ Proactive; Mitigation/ Proactive	0
11	Other, internal projects may benefit from the application's feature set, so the project is subject to those requirements being defined	Significant delay and additional work in integrating other requirements into the project	EMV = +$300,000; note: this is notional value of the benefit to the other projects.	Negotiate with Project Management Director and other project managers to share in the development costs. If not, then ask that any integration be borne by the other projects.	Share/Proactive; Transfer/ Response	0
			$397,500	Approx. $360,000	NA	Approx. $37,500

A good part of the development of risk responses is looking at the options and at the cost benefit of each and seeing whether one or a combination of the responses may be most effective in addressing the risk.

Now, let's review some of the risks and treatments given above to explore the different types of responses.

- **Risk 1. Delay in approval for the project start.** If we know based on past projects that approvals have reduced the time, we have to execute the project, then we can estimate what those delays have been on average and add that time to our schedule. The other action we can take is to discuss with legal counsel and financial managers on both sides to see what has held up contract approvals and allocation of funds. This latter option requires relationship building and negotiating skills.

- **Risk 4. Emergency incident causes work delays.** There can be a wide range of different types of emergency incidents, with different time delays depending on the response involved, so identifying each and the extent might be difficult to factor into the schedule ahead of time. There is also a potential for it to change the scope based on new requirements discovered during the response. The delay is addressed through two methods of risk transfer: contractual risk transfer through a clause in the contract that allows for a no-cost extension in the event of an emergency incident. The second is business interruption insurance, which some companies carry to address extended outages and time for recovery. The last impact is one that is addressed through mitigation by negotiating any scope changes.

- **Risk 11. Other, internal projects may benefit from the application's feature set, so the project is subject to those requirements being defined.** The impact of this type of risk presents both positive and negative risks. The option we take is to negotiate with the other project managers and Project Management Director to share in the development costs. The other option is to delay any potential changes and ask that integration costs be borne by the other projects, a method of mitigating the risk.

Let's look at an example of avoiding a risk. When the project team was in the early stages of identifying requirements and the client team asked that the application work on mobile devices. The requirement was analyzed and deemed to create several issues: a security risk, mobile device requirements were difficult to define (with several different platforms), and the work would also require significant development time and testing. The option was chosen to avoid the risk and to exclude this requirement from this project. As for acceptance, some of the risks that were lower on the risk prioritization matrix (risks 15–20) would be accepted as they could be managed through the project planning process and did not pose a significant impact to the project objectives.

A risk register is not a risk plan, so after analyzing our risks, identifying, and developing responses, we want to pull this together into a plan that explains the strategy, methods, responsibilities, and reporting and management mechanism, as well as communications and tools employed to manage project risks. The items below describe each area of a narrative risk plan, followed by an example in Table 10.12 based on the project example in this chapter.

Table 10.12 Risk Management Plan

Risk Strategy	Our software development company approaches this project with a risk averse attitude and takes a proactive approach to address risks, seeking to transfer risk through contractual means and managing expectations and adhering to IT and government regulatory requirements. We have developed similar projects but feel that the greatest degree of risk lies with: • Subcontractor for eLearning courses failing to deliver on-time or according to quality requirements. • The computer and server configuration used by the end-users that will deliver the application may prove not to support the application sufficiently. • Other City agencies that want input into the development process.
Risk Organization, Management, and Reporting	*Roles and Responsibilities:* The Project Manager, Lead Software Developer, and Quality Manager will monitor and manage the risk plan. Each team member may be assigned responsibility for specific risks. Milestone reviews will include risk reporting and consultation on any active risk responses or new risks surfaced with the internal sponsor and Client POC. *Risk Response Management and Reporting:* • The project team estimates that 20% of the budget will be devoted to managing risk, including schedule and other project elements, at the start of the project and will be adjusted during the middle and end phases of development. • The risk plan will be reviewed and actively managed once a week along with the overall project plan, and at key milestone reviews. • The Risk Report will include the use of a Risk Planning Board, with risks drawn from the Risk Register. The Risk Register will be updated as risks are addressed or avoided during the development process. • The Risk Report will first be reviewed by the Project Team. The Project Manager will discuss any risks that may require escalation with the Project Sponsor, and ultimately with the Client POC that cannot be addressed internally.
Risk Management Methodology	**Risk Identification** We will employ the use of a Risk Breakdown Structure to capture risks and use the following risk identification techniques for identifying risks: • Conduct and organizational risk scan: strategic risks, financial risks, operational risks, and hazard risks (including threats). This will include organizational historical records, including previous projects. • Review project plan elements: scope, schedule, budget, quality, procurement, human resources, stakeholder and communications. • Brainstorming sessions with key stakeholders, including project team members, client team, and key vendors to elicit their experience with risk. *Risk Assessment and Analysis* We will employ the use of a Risk Rating Matrix to assess and prioritize risks. We will use a Risk Register to account for risks and the following methods will be used to analyze risks: • Conduct an initial EMV analysis for probability and impact analysis with a team discussion. • Bowtie Analysis, conducted as a facilitated discussion with key team members and stakeholder for each risk event they participate in. • A more in-depth fault tree and event tree will be used for risks that can impact the budget or schedule with a variance greater than 5%. • A decision-tree method may be used to evaluate options such as outsourcing vs. in-house staff employment. • Early quality testing will simulate the computer working environment to address end-user configuration risk. *Risk Response* Prioritized risks, with the potential for 5% or greater variance from the budget or schedule, will be addressed through risk transfer or mitigation methods. Mitigation methods will employ rigorous project planning methods: quality management, procurement methods, human resources management, and stakeholder and communications management, and risks that are not prioritized will be reviewed the ensure that they are addressed as part of this project planning process. Where possible, we will look to avoid residual risks that exceed the 5% variance threshold by avoiding certain options or establishing a contingency fund to reduce residual risks to 0.

(Continued)

Table 10.12   (Cont.)

Risk Thresholds and Definitions	*Risk Definition:* An event that can negatively or positively impact the accomplishment of the project's goal and objectives (deliverables). Risks will warrant a risk review that may impact quality of deliverables or outcomes requiring significant rework that would exceed the variance thresholds.
	Variance Thresholds: 5% variance of project budget and 10% variance of project schedule will trigger a review by the sponsor and senior stakeholders.
	Showstopper Criteria: • Product completely fails a critical test, requiring a return to a previous development stage. • Customer does not accept the product at a key milestone. • Key stakeholder seeks to change product requirements during the development stage.
	Any team member, the Client POC or internal sponsor may identify a showstopper, but it must be recognized by the Project Manager and Client POC. Once recognized, the project team will call for a full project review and develop recommended courses of action to address if not addressed in the risk register.
Communications and Tools	The Risk Plan as laid out here will be integrated into the project plan and will be reviewed with the other plan elements (HR, quality, etc.). The software tools that are used to track risks below will be updated on a regular basis as part of the planning and plan management activities.
	An online spreadsheet will be used as a risk register to account for all risks. A risk planning board will be employed to actively manage risks during the project execution phase, with all final updates to the risk register.

- **Risk strategy:** Describes the organization's risk attitude for this project in comparison with other projects, the organizational approach to project risk management, and what are critical risks to the project.
- **Risk management organization, management, and reporting:** Identifies the main project team members who will manage risks. It will also identify any stakeholders that will be closely consulted with or help to manage project risks. The active risk response and management explains how much of the budget will be devoted to project risks, and how this will be managed during the course of the project, as well as frequency of a risk plan review, what tools might be used, and the reporting structure.
- **Risk management methodology:** Explains the how risks will be identified, assessed, and analyzed and responses developed, and what criteria might be used for those responses.
- **Risk thresholds and definitions:** Identifies the measurable project parameters for variance thresholds that may trigger a risk and review by the risk management team and stakeholders. It also defines terms for a risk event and a showstopper risk. It defines who is authorized to raise a showstopper risk and how it will be addressed.
- **Communications and tools:** The last section describes how the risk plan relates to other project plans and activities. It also indicates any tools, such as software applications that will be used to manage risks.

When writing and presenting a risk plan, it may make sense to include a narrative portion and then spreadsheets and software tool reports (snapshots etc.) in separate files, as a word-processor document may not work well in combination. For example, a risk

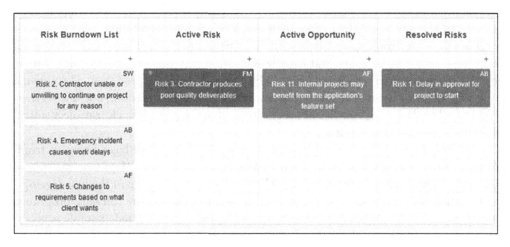

Risk Burndown List	Active Risk	Active Opportunity	Resolved Risks
+	+	+	+
SW Risk 2. Contractor unable or unwilling to continue on project for any reason	FM Risk 3. Contractor produces poor quality deliverables	AF Risk 11. Internal projects may benefit from the application's feature set	AB Risk 1. Delay in approval for project to start
AB Risk 4. Emergency incident causes work delays			
AF Risk 5. Changes to requirements based on what client wants			

Figure 10.8 Risk planning dashboard.

register in a spreadsheet (MS Excel or Google Sheets) may contain too many columns to appear in a Word or Google document.

MANAGE RISK PLAN: MONITOR AND ADDRESS RISK EVENTS

Once a risk plan is developed, it is a matter of implementing it. Plans are tracked and reported, with status of deliverables, any variances from the plan (primarily cost and schedule) or deviations (for example quality) and risks monitored. The plan, including the methods, is reviewed to ensure they are effective. New risks may be discovered, analyzed, and added to the risk register, while some individual risks may be recorded as no longer valid as the project develops and makes them invalid.

While there are a number of risk management systems (RMS) and applications for managing projects, ultimately the risk register, or its equivalent (usually an electronic file) where risks are tracked, becomes the core tool for actively managing project risks. When risks are reviewed, their status is updated, including any active response implementation and outcomes. This may track how much contingency reserve, by budget or schedule, is available at the end of a project, the more comprehensive element that captures the evaluation of the risk plan, and risks that were encountered and addressed culminates in the lessons learned, sometime referred to as an After-Action Review, which informs how the project risk management may be enhanced for future projects.

DRAWBACKS OF A RISK REGISTER VS. RISK PLANNING BOARD

I want to address some of the limitations of a traditional risk register and its use in managing project risks. The traditional method of using risk register may be fine but often is heavy on methodology and light on active problem solving. If you have someone who is actively managing that risk register or risk mgt. software on a regular basis, then it may work well to manage risks. More often than not, a risk register is a canned plan that usually resides in an electronic spreadsheet with many fields on it, is typically highly technical, not updated on a regular basis, and usually goes stale very quickly after it's developed. These risk plans are often not shared or available to those who may need them. Documents that sit in a folder or isolated software repository are not living breathing documents that are part of a live dashboard that makes them front of mind, providing clear project status and decision-making opportunities.

One method that I find effective in managing risks on an active basis borrows from agile project management methods, the planning board or Kanban (Japanese for "Card You Can See", basically a bulletin board). The Kanban is used in one of the key development tracking tools and is in wide use in tech. companies, with many software programs that offer the agile development functionality and features. An adapted example is given in Figure 10.8 illustrating risks that are on a burndown list.

A risk planning board based on the Kanban offers greater visibility and agility in being able to find and drill down into the details quickly and integrate with other documents or dashboards on the project. In the sample in Figure 10.8, initials indicate the team member responsible for the risk and can be color coded to indicate its status and type. Many systems integrated with office productivity and organization knowledge management software, such as Office 365 and Teams or Google Workspace, and offer the functionality of attaching files, assigning responsibility, and tracking progress within the risk on an ongoing basis. Creating custom reports is far easier than with the use of a spreadsheet, and being able to sort and quickly move from risk to risk, or by person responsible, makes reporting much more dynamic. Project risk plans and risk management do not happen in isolation, so other plans need to be reviewed and updated in accordance.

NOTES

1 The term *Ecological* is used instead of *Environmental* as the latter term is used in this book and commonly elsewhere to describe a general and ongoing situation in the industry or economy.
2 Enterprise Risk Management (ERM) is a risk management approach to managing the potential impacts from risk that can influence the accomplishment of the organization's strategic objectives that drive shareholder/stakeholder value. This differs from traditional risk management which is loss focused and designed to protect the organizations balance sheet and profit/loss goals. ERM is a much more global look at risk.
3 The New England Hurricane of 1938, often referred to as the Long Island Express. While it may landfall in Bellport, NY in the middle of Long Island, it was the equivalent to a cat.3 hurricane with sustained windspeeds of 120 MPH, causing major damage in the New York City area as well. Pierce, C.H. (1939). The Meteorological History of the New England Hurricane of September 21, 1938. Monthly Weather Review, 67, 237–285.
4 New York City Emergency Management Risk and Recovery Unit (2014), Hazard Mitigation Plan; NYC Risk Landscape Ch. 4.1, p. 51.
5 Samenow, J. (2017, September 22). 60 in. of rain fell from Hurricane Harvey in Texas, shattering U.S. storm record. The Washington Post. See: www.washingtonpost.com/news/capital-weather-gang/wp/2017/08/29/harvey-marks-the-most-extreme-rain-event-in-u-s-history/
6 Program Evaluation and Review Technique (PERT) methodology was developed by the US Military in the late 1950s to account for uncertainty and risk on projects, and in its early stages,

it used three-point estimates for tasks in its analysis. Source: *Kendrick, T. (2015). Identifying and Managing Project Risk: Essential tools for Failure-Proofing your Project. American Management Association.*

7 BIM and GIS Integration: Models and Maps Working Together www.esri.com/en-us/industries/aec/overview/gis-and-bim

Chapter 11

Developing the Stakeholder and Communications Plan

Tell me and I forget, teach me and I may remember, involve me and I learn.

Xun Kuang, "The Teachings of the Ru"

I keep six honest serving men (they taught me all I knew); Their names are What and Why and When and How and Where and Who.

Rudyard Kipling

STAKEHOLDER AND COMMUNICATION MANAGEMENT

In emergency response, the key responsibility in effective communication is "getting the right information to the right people at the right time, so they can make the right decisions".[1] In the same manner, communication of project plans, whether presenting it to the client and sponsors for project approval (including for proposals), status reports, or for project closure, all center around decision-making. As discussed in Chapter 5 on scope and stakeholders, we define and assess stakeholders in the project to get "buy-in" at the start of the project then effectively manage their expectations throughout the project lifecycle. Each group of stakeholders has a different level of interest and influence depending on their authority and relationship to the project, so in managing communications we need to craft our messaging and channels to suit each stakeholder's needs.

The first step in planning project communications starts with identifying the target audience, stakeholders, and their communication requirements. A Project Communications Plan Matrix shown in Table 11.1 provides a format for defining these requirements. We will continue with the sample project that we used in the previous chapter; the software application for managing registration for special medical needs client records.

How would we set up communications on a nine-month project of this type (software application development)? Using the Project Communications Plan Matrix, we would assemble the following:

- Project contact list for major Point of Contacts (POCs) for each stakeholder group, with contact info. (email, phone/cell number, etc.). This may include creating several group email lists.

DOI: 10.1201/9781003201557-13

Table 11.1 Project Communications Plan Matrix

Target Audience	Comms. Objective	Messaging	Frequency	Channels	Provider
Project Sponsor (Project Management Director)	Maintain project support and access to resources	Status of project deliverables, milestones, and budget; critical risks	Monthly	Written (product dashboard) and oral report	Project Manager
Project Client (Director of LCEMA)	Meet expectations and support for project	Status of project objectives, milestones; critical risks	Biweekly	Written (project dashboard and oral report, project status meetings	Project Manager
Project End-Users (ARC, MRC, and other volunteers)	Meet product requirements	Status of product deliverables; development and testing milestones	As needed, more frequent once product is ready for testing	Product deliverable status report, email, and dashboard	Lead Software Developer
Project Team	Committed to project objectives and schedule	Status of schedule, deliverables and tasks, risks and quality	Minimum weekly (some daily)	Face-to-face (F2F) meetings, in-person team development meetings, and project dashboard, email for task assignments	Project Manager
Vendors/Subcontractors	On target with scope and schedule	Vendor tasks, schedule, quality feedback, payments	Biweekly/weekly (in contract)	Vendor status meetings and email	Vendors
Liberty City Social Service Agencies (Senior Affairs Dept., Dept. for People Experiencing Homelessness)	Maintain project support and agreement with product outcome	Status of product deliverables, development and testing milestones	Monthly	Product deliverables status report, email	Project Manager/Project Client
Liberty City Legal Dept.	Maintain project support and eliminate showstoppers	Project risks specific to legal/regulatory matters (cyber sec. etc.)	Monthly (more frequent at the start)	Product deliverables status report, email	Project Manager/Project Client
Liberty City Info. Tech. Dept.	Maintain project support and eliminate showstoppers	Project risks specific to IT and cyber sec.	Monthly (more frequent at the end)	Product deliverables status report, email	Project Manager/Project Client

- File repository for all project planning and development documents. The common systems used these days are often shared folders of which there are many versions (at the risk of this becoming outdated: OneDrive (MS), SharePoint (MS), Dropbox, Box, Google Drive). There are also project planning platforms, such as Basecamp, Monday, and Atlassian, which offer project scheduling, information management, and file sharing, as well as other functions. This would include file sharing permissions appropriate for stakeholder needs. The following is a folder system, with file content descriptions, that I use on projects, which borrows from the PMBOK© Project Management Knowledge Groups:[2]

  - **1.0 Legal and contract documents:** This includes all contract documents, including any RFPs, bid documents, proposals, contract documents, and any contract amendments.

  - **2.0 Delivery:** This contains documents covering instructional delivery when that is included, as well as any projects that require any special or extensive shipping, transportation, and assembly.

  - **3.0 Product development:** This includes all files related to product development: blueprints, design documents, outlines, process maps, product reviews, pilot tests, testing results, etc. When using a waterfall development cycle, I further break this out as follows:

    - **Design Documents:** Design documents are comprehensive plans that describe the key features and guidelines for building a product or system, acting as a reference throughout the development process. This may include outlines, blueprints, and a list of product requirements.

    - **Alpha (first prototype):** These files are an early working version of the product used for product review or walk-through.

    - **Beta (second version):** This is a version fit for end-user testing.

    - **Gamma (third, final version):** This is a final version ready for launch or production.

    - On some projects there may be subsequent development stages such as a Delta version, after the product has been in production for period of time and further updates are planned. Also, note that some organizations use different development stages, similar to those above, such as Bronze, Silver, and Gold; or Strawman, Woodenman, Tinman, Ironman, Steelman, terms used originally in the US military.[3]

  - **4.0 Integration and change management:** Documents related to change control processes and procedures, any change requests, and status of changes. This can include any documents that integrate updates between component plans, e.g. risk to quality and HR to cost.

  - **5.0 Scope:** Documents covering the project scope, including scope statement, WBS, and a preliminary set of product requirements (see quality).

  - **6.0 Time management (schedule):** Project schedule and updates. Note – often schedules may be managed using online systems (MS Project, Asana, Atlassian [Jira], and Smartsheet, including those referenced above).

  - **7.0 Cost management (budget):** Project cost estimates and budget, and ongoing updates.

- **8.0 Quality management:** Product quality documents, quality plan, quality requirement list, quality register, as well as any documents concerning product pilots, product testing, and test results.
- **9.0 HR management:** Human resource documents such as the project org. chart, client org. chart, hiring documents, staff contracts, performance reviews, and professional development records.
- **10.0 Communications:** Contact list, project status meeting agendas and notes (if they pertain to specific plans, such as quality and risk, they may be placed in those folders).
- **11.0 Risk management:** Project risk management documents, including the risk plan, risk register, as well as any documents concerning risk responses.
- **12.0 Procurement:** All procurement plans, procurement guidelines and procedures, RFPs for contracted products/vendor services, subcontracts, vendor contracts, and correspondence with contractors/vendors. Note – subcontracted product and service quality evaluations may be stored in the Quality Management folder.
- **13.0 Stakeholder management:** Documents covering the stakeholder plan, stakeholder analysis matrix, meeting rules, stakeholder engagement rules, and correspondence related to stakeholders that falls outside of the folders above.

- Schedule all major meetings: Kick-off, status meetings, milestone/stage gate reviews, including any product walk-throughs, pilot and user tests. Some meetings, aside from status checks, that may take place well into the future, for example beyond six months, might be scheduled at three months out when schedules may be more clearly defined. Meetings, as discussed earlier in the chapter, need to be effective, so any status meetings that take place close to reviews or other major meetings may be dispensed with. Use the RAM to account for who needs to attend which meetings, and make sure to account for holidays, vacations, or organizational events (strategy meetings, outings, etc.) when setting the schedule. Lastly, ensure you have a system for meeting preparation, setting and sending meeting agendas, read ahead materials, and how you will record meeting notes for follow-up actions.
- For virtual meetings (using platforms like MS Teams, Zoom, Google Meet) make sure that all participants are able to access the platform, i.e. have an account and are able to attend beyond their organization's fire-wall, and know how to log in, etc. Have a set of virtual meeting rules, whether you plan to record meetings for participants unable to attend or for meeting records. Some state laws and organizations prohibit recording virtual meetings unless everyone attending has granted permissions. Also, make sure you have a plan for virtual meeting features, including someone who will manage the chat if this is not the meeting facilitator, when participants have their microphone and camera on, and who can share screens. See presentation recommendations in the next section.

As we started off this section, we want to meet the expectations of our primary stakeholders who have the greatest influence and interest in the project. If we deem the project sponsor, client (Point of Contact or POC), and project team as priorities, then we can best manage those expectations by ensuring that we communicate efficiently and effectively. First of all, this means choosing the channel or format:

- Phone or virtual (real-time) call one on one or conf. call with a group; the latter is usually planned; the former can be impromptu.
- Instant messaging (Text, Teams chat, Slack, Signal, etc.).
- Email, usually with key messages and an attached project report or document.
- Project dashboard (as described above).

The choice of channel will largely be driven by the frequency of the communication, the stakeholder's preference, and the urgency in delivering the message to the recipient.

As Mark Twain once wrote "I didn't have time to write a short letter, so I wrote a long one instead". Crafting a message is an art that takes time to master. Luckily, there is an abundance of communications courses available at FEMA's independent study portal (online learning platform) that can help, and even more available through a variety of other sources. The most critical part of messaging is identifying what action you want the recipient to take. Similar to any situation assessment and next steps to follow, the basic framework is as follows:

- What is the situation status?
- Is it urgent, important, or can wait? Is there a time frame to respond?
- Is there an impediment to or opportunity for the project (achieving an objective, quality issue, or risk)?
- What do you need the person or group to do? This is the call to action or for a special meeting; or perhaps a heads-up for the next meeting.

As communication is part art, cultural competency, organizational, community, and country-based, including linguistic differences, should be taken into consideration. Some organizations and countries are more formal or less formal with communications. Awareness also needs to be paid to the organizational and interorganizational power dynamics and relationships in handling communications. The subtleties of who may need to be brought in on specific decisions and when can take time to learn depending on the project managers' familiarity with the more senior level management, both internal and external to the organization. Lastly, communication is a two-way process, so active listening skills and feedback are important to take into consideration as part of the process.

CULTURAL COMPETENCY, DIVERSITY, EQUITY, AND INCLUSION

From my years of working overseas, one of the key lessons learned was the need to better understand the culture in which I was working before deciding on any major course of action. My preferred steps before embarking on a new project in a new country or area were most often to learn the language, read the well-known authors, study the history, and take part in the cultural life of the country. This often included many meals with the local wines or alcohol, numerous toasts, and much dialogue. My intent was to absorb the culture and gain a better respect and understanding of the core values, beliefs, assumptions, and attitudes, and what might motivate people, or alternatively demotivate people, to behave in certain ways.

Getting a better understanding of a culture that we may be unfamiliar with, whether in another country, another region, or with another community, requires us to recognize our own core values, beliefs, assumptions, and attitudes, and to acknowledge that there may be key differences in how we operate vs. how people from a different culture may operate. I recall one of my first meetings in Croatia, with the Minister of Health to find out more about the health needs of the refugee and internally displaced persons (IDPs) population. I brought along my Admin. Assistant who served as an interpreter since I was not able to speak the language at the time. During introductions, he asked if there was anything he could get me. I was so task oriented and ready to get down to business that I presented him with a request for a litany of items concerning health statistics on refugees and IDPs. After waiting for me to finish with kind attention, he smiled and the interpreter explained that he asked me if he could get me some tea, coffee, or juice. Lesson learned: he was focused on the relationship and treating me as a guest, and not just on the task.

We need to be careful to distinguish between cultural norms and stereotypes. Stereotypes are set perceptions and opinions which are usually oversimplified. We know that within a culture there are wide disparities of behaviors driven by values, beliefs, and attitudes, as well as the historical and current social situation. Keeping this in mind, developing a relationship first through active listening and building trust is the first step. Understanding the social dynamics can help build cultural competency, such as figuring out whether the culture puts individuals or the community first; power is drawn from a person's position or the community tends toward equality; risk tolerance is conservative vs. daring; communication is direct vs. indirect; or there is a strong task orientation vs. relationship building.[4] While we should be aware of cultural norms, we should be careful not to paint people with a broad brush and assume these values and behaviors are general points of understanding about a culture.

Emergency management has been facing several key issues related to diversity, equity, and inclusion (DEI). While the specific landscape may have evolved as of this writing, these challenges and ways that they might be addressed within project management communication include the following:

- **Representation and leadership:** Lack of diversity in leadership positions within emergency management agencies can hinder understanding and addressing the needs of diverse communities during disaster planning. Increasing representation of underrepresented groups in leadership and project team roles is crucial for more inclusive decision-making.
- **Access to resources:** Historically marginalized communities may have limited access to resources and information which can make it more challenging for them to adequately prepare for, respond to, and recover from emergencies. Ensuring equitable distribution of resources and information is essential, so making sure that project requirements take this into consideration is key.
- **Language and cultural barriers:** Language barriers can hinder effective communication and understanding of emergency procedures and safety measures. Additionally, cultural differences may influence how individuals and communities perceive and respond to emergencies.
- **Community engagement:** Effective emergency management relies on community engagement and collaboration. Some communities, particularly those marginalized

or historically mistreated, may be hesitant to engage with emergency management agencies due to past experiences and mistrust. Making the sincere, concerted, and committed effort at outreach when a project calls for it can help engage the community. That means going out to the community, not always having them come to you.

- **Inclusive planning:** Emergency management plans and policies need to consider the unique needs and vulnerabilities of diverse populations. Inclusive planning involves tailoring strategies to address the specific challenges faced by different communities.
- **Intersectionality:** Recognizing that individuals belong to multiple identity groups, e.g. race, religious, gender, ethnicity, disability, is critical for understanding their distinct experiences and needs during emergencies.
- **Training and sensitivity:** Emergency management personnel should receive training in cultural competency, unconscious bias, and sensitivity to avoid unintentional discrimination or prejudice during emergency responses.
- **Data collection and analysis:** Collecting demographic data before and during emergencies can help identify disparities in planning for disaster impacts and response. However, inadequate data collection on diversity-related metrics may hinder efforts to address disparities.
- **Inclusive communication:** Communication needs to be in accessible formats, multiple and culturally appropriate languages, and through diverse channels and messengers to ensure that everyone can receive and understand critical project-based and planning communication when it is addressed to the public.
- **Trauma-informed approach:** Recognizing and understanding the historical and ongoing trauma experienced by certain communities can inform planning for emergency response efforts that are sensitive to their unique needs.

Addressing these DEI issues within projects undertaken in emergency management and public safety is essential for enhancing resilience in communities and ensuring that no one is ignored by planning efforts and disproportionately affected by the impacts of disasters or emergencies. The mantra "nothing about us, without us" is important to bear in mind in any major planning effort. By actively promoting DEI, emergency managers can build more effective, responsive, and compassionate systems that serve all individuals and communities equitably.

PRESENTATIONS

Project managers, at some point, are asked to present their proposals, plans, or progress reports, whether to the client, senior managers, the team, or to other major stakeholders. Good presentation skills make for effective communication. We have all sat through a boring presentation, wishing we could hit the eject button, or perhaps fain illness in order to escape. What makes for a boring presentation? I will list just some key mistakes:

- a presentation that contains information that is not targeted to the listening audience;
- presentation visuals (slides, mainly) that are too dense: font is too small, text heavy, contain spreadsheets or complicated data that is difficult to read, or just have too many slides;

- presenter reads from the slides verbatim, sometimes without looking up at the audience to make eye contact, or worse turns their back on the audience to read from the screen (Ugh!);
- presenter delivers in a virtual monotone, does not vary their pace – either too slow to too fast – and does not engage the audience in an effective way (more on this later).

On the other hand, giving presentations does not always come naturally to some people. There are other general aspects of poor presentations, such as failing to rehearse the presentation, apologizing too often, or using terminology and acronyms that the audience does not understand. We could go on about what makes for a painful, live presentation; however, I want to address what can make for effective live presentations. Here are some of the top actions to take when developing an effective presentation:

1. Identify the target audience for the presentation and what they are interested in. Consider what some of their main questions might come to mind about the project, besides the basics (scope, schedule, and cost):
   - Why are we undertaking this project? What problem does project solve?
   - What are we going to gain? What are the benefits? What is in it for them (the audience)?
   - What are their concerns and how are these addressed in the project plan?
2. Write out the main points or, if you are less practiced, a script, but don't read from it. Practice the presentation, so you are familiar with the contents. The audio delivery is the key part, and slides/visuals are secondary. Always be prepared to "stand and deliver", even without slides.
3. Keep in mind that when delivering a presentation in-person (to a lesser extent online), a majority of spoken communications come from your body language in-line with the 55-38-7 rule[5]:
   - Overall, 55% of the message comes from body language: appearance, posture, facial expression, and hand and head movements, and eye contact with the audience.
   - Overall, 38% of the message comes from your voice (the way the presenter speaks): pitch, pace, and projection.
   - Overall, 7% of the message comes from the actual words or content delivered.
   - It helps to practice in front of a video camera or an audience, especially if you may be nervous when speaking publicly. For those who are nervous or are looking to improve their skills, Toastmasters International is a great non-profit organization that helps people "practice public speaking, improve your communication, as well as build their leadership skills".[6]
4. Engage the audience emotionally. This can be done by the following:
   - Using imagery to convey the message, as the saying goes "a picture tells a thousand words". More images and less text are often a better way to go.
   - Telling a relevant story that relates to the problem that the project is supposed to solve, but keep it short enough to maintain the listener's attention. A good model is to start with something particular, using a story which connects to the project's purpose, then goes to a general set of points such as the fundamentals around the project, then perhaps comes back to the particular or a story at various intervals.

Using a story can engage the audience, as the audience likes to follow what is happening with the plot and to hear how the story ends.

- Ask direct questions of the audience if the opportunity allows. When I am invited to speak to an audience on topics concerning emergency management, the one question I often ask is "Do you have a Go-Bag and an emergency plan for your household?", and then ask for a show of hands to take a poll on the responses, then I tie it back to the talk. This can help bring your audience into the presentation at the start.
- Be enthusiastic and use humor or give certain points an emphasis with feeling, but don't overdo it and come across as being overly dramatic. Remember it is a project presentation and not a Broadway show!

5. For slides/visuals make them effective by following some of these rules:
   - As stated above, use images instead of text when possible or complement the on-screen text (OST). Make sure the visuals communicate the message and are easy to decipher.
   - If using OST, summarize your points and keep the OST as brief as possible with short phrases. Use the 6 × 6 (or 7 × 7) rule: six words per line by six bulleted lines of text. This rule can be stretched at times, but do your best to stick to it for greater effectiveness. People read faster than they can listen; so when given a choice, the audience will either listen to the speaker or read the OST, but not both, so it creates cognitive dissonance and can break the presentation's flow.
   - Don't create too many unnecessary slides that do not serve a purpose. If the rules are followed above, it generally takes between 3 and 5 min to cover each slide with content you plan to deliver, not including the title, ending, and transition slides. The number of slides for a 20-min presentation should typically be between six and eight slides long, not much more beyond that.
   - Make the OST font large enough for the audience to read. Don't present them with an eye examination! In general, a 20-point font size or higher is readable from a distance (this may differ for online presentations). If you are delivering in-person, I recommend going to the very back of the room and seeing if you can read the OST from that distance.
   - For detailed information, such as schedules, and budgets, summarize the message on the slide, such as showing milestone dates on a timeline or summary budgets for the main deliverables. Use handouts for the audience to cover what can't be displayed on the screen.

6. Master the technical mechanics. Make sure that you are familiar with the software for visuals (PowerPoint, Google Slides, Keynote, etc.) and that the screen projector is working beforehand and the presentation is accessible from the computer, so you do not panic at the moment when you have to deliver. Give yourself enough time to do a system check before the audience arrives. If delivering online, then make sure that you know how to work the controls of the virtual meeting platform, including the use of the chat feature, microphone, and your on-camera video (eliminate unnecessary background audio noise and video background visuals).

7. I recommend avoiding the use of slide animations for OST, images, or transitions, unless you are experienced as a presenter. Animations can be distracting, time

consuming, and appear clunky when not well practiced. Frankly, they can be quite annoying when used too much.

8. Manage your time, making sure to start and end on time, as well as anticipate audience members arriving late or leaving early by placing the main messages in the middle of your presentation.

If you are looking for more tips, I recommend watching how the best presenters deliver their talks (online videos are a great source) and generously borrowing from them. With time and practice, presentation skills can be developed to make communication more effective and presentations less of a chore and more of a core competency.

NOTES

1 FEMA, US Department of Homeland Security (Feb. 2014). *IS-242.B. Effective Communication – Student Manual*, p. 3.14.
2 Project Management Institute. (2017). *A Guide to the Project Management Body of Knowledge* (PMBOK guide). p. 25.
3 Lucas, H. C. (1981). *Implementation: The key to Successful Information Systems*. Columbia University Press. pp. 94–95.
4 Ng, Mui Hwa. Motivating Across Cultures: Encouraging Higher Performance 22 April 2021. Aperian Global.
5 Sometimes referred to as the Mehrabian rule after Professor Albert Mehrabian. Mehrabian, A. (1977). Nonverbal Communication. United States: Taylor & Francis.
6 I have been a member of Toastmasters during my career and have found it enormously beneficial, as well as a great place to meet new people in a supportive environment. See www.toastmasters.org/

Chapter 12

Procurement, Contract Management, and Reporting

External procurement of equipment, materiel, supplies, and/or services is almost always required on all projects, and, depending on the client agency and internal organizational procurement rules and requirements, requires significant time and attention to ensure that it delivers what is required. This may require a "make or buy" decision for a product/service. This chapter is intended to provide better practices to meet the planning needs of organizations that are submitting responses to public request for proposals (RFPs), request for quotes (RFQs), bid requests, etc. and managing procurements throughout the project lifecycle but may also help in crafting the same. It also is limited to a general set of guidelines and practices, as there are many details, requirements and variations, especially by different industries, first response agencies, and by governmental type and level.

Much like the overall project planning process, project procurement involves the following major processes[1]:

- Plan procurements and contracting process
- Request responses from and select vendors/contractors
- Manage contract performance and close out contract

PLAN PROCUREMENTS AND CONTRACTING PROCESS: GOVERNMENT GRANT FUNDING

A considerable number of projects undertaken in emergency and public safety in the United States are funded by government grants, most of these at the Federal level, but also undertaken at the state and local level, or when Federal funding requires a percentage of cost sharing at the state or local level. In fiscal year (FY) 2021, for example, Congress allocated $1.06 billion for FEMA's 1 Assistance (HMA) programs, which includes the Hazard Mitigation Grant Program, Pre-disaster Mitigation program, and Flood Mitigation program and other similar programs mitigation programs.[2] In FY 2022 FEMA's budget request for its "Disaster Relief Fund" was $14.9 billion which include emergency preparedness and response, recovery, and mitigation efforts.[3] In general, the greater the funding for a project and more complex it is, such as large-scale engineering projects, the more likely government grants are supporting it.

DOI: 10.1201/9781003201557-14

Governmental agencies' mission in procurement is to obtain the best value for the tax-paying money, while maintaining an open, fair, and an equitable process.[4] Toward this end, quite often grant guidelines for public agency procurement may include additional stipulations beyond those found in a standard commercial contract, including required forms and documentation to address:

- Potential, actual, or appearance of conflicts of interest.
- Where goods may be sourced from or the percent of the product produced domestically, e.g. US Federal "Made in America" rules.
- Preference for business location and ownership, e.g. requirements for equity in competition such as favoring small businesses; minority, women, veteran owned businesses. Some of these requirements are intended to meet social and economic equity goals and may include certain "set-aside" amounts for disadvantaged businesses.
- Americans with Disabilities Act (ADA) compliance provisions.
- Diversity, equity, and inclusion staffing and hiring requirements.
- Paying staff prevailing wages.
- Where and how the work may be performed, whether on-site in the agency's location, home state, or in the national boundaries, with potential limitations for any outsourcing to foreign countries.

These types of procurement stipulations may differ based on the level of government (National/Federal, County, Tribal, Local) and geographic location, as well as legal framework and changes in political winds. Organizations, whether governmental, for-profit, or non-profit, will have a procurement department or unit (may be part of the Finance and Admin. Dept.) with their own set of purchasing guidelines, rules, and practices, and they will often help to shepherd the procurement process.

PROCUREMENT MECHANISMS AND CONTRACT VEHICLES

Different types of procurement mechanisms are needed depending on the type of product or service procured. Table 12.1 describes the differences in procurement mechanisms and where they are most suitable.

There are also subsets of the public RFP or bid request process that identify a group of trusted contractors/vendors from which products and services can be procured over a longer term, as part of a larger program, or standby contracts when goods or services are needed on an urgent basis during an emergency response. This is done to reduce the administrative work and lengthy procurement process. These include as follows:

- **Indefinite delivery/indefinite quantity (IDIQ):** The IDIQ process defines a range of products and services that a contracting agency (client) is seeking, prequalifying a set of contractors from which to choose the goods and services for specific scopes of work.[5] The contracting agency executes an IDIQ Master Contract with qualified contractors. Once contractors have been defined, a project is defined as a Task Order, sometimes referred to as a "Job Order", a smaller project by cost and schedule. The contracting agency can decide to solicit proposals from a smaller set of contractors or all qualified contractors for the scope of work. The task order defines the scope,

Table 12.1 Procurement Mechanisms

Procurement Mechanism	Purpose	Example of Product/Services
Request for information (RFI)	Used when identifying suitable vendors/contractors, or when looking to find out more about a specific product or service for products are services new to the client organization. An RFI may be used prior to an RFQ, RFP, or bid request.	Can be used for a variety of products or services.
Request for quote (RFQ)	Most often used for standard goods, services, or commodities, where product and service specifications and performance are well defined; or to obtain competitive pricing from potentially new vendors/contractors. May also be used for fulfilling ongoing needs.	Computer equipment, printing, cloud computing services, disposal of post-disaster debris.
Request for proposal (RFP)	RFPs usually include a statement of work that defines what key requirements the client organization is seeking to meet. Often for products and/or services which are not standard and where more information may be needed as to how the contractor might approach the statement of work and there may be variation in the quality and cost.	Plan development; training course development and delivery (eLearning or classroom based); emergency exercises; complex computer applications and technology, e.g. ICS management software, cybersecurity assessment, and body cameras for public safety.
Bid request (internationally often referred to as a Tender)	Similar to an RFQ, but used for large-scale, complex projects, such as construction, engineering, turn-key systems. Often used in conjunction for "Design-Build" projects. This is typically conducted as an official public procurement, which is advertised, and the process is tightly regulated by the jurisdiction.	Constructing berms, IT backup sites, building a law enforcement or fire service training center, establishing an EOC.

schedule, and cost that serves much like a purchase order, intended to have shorter turnaround process. Once the task order is executed between the parties, the IDIQ Master Contract governs the contractual terms, so instead of months or longer for an RFP and contracting processes, the agreement for the project can be in place in weeks.

- **Unit price contract:** Unit price contracts are a type of fixed price contract (see contract types below). It specifies a fixed price per standard unit of measurement, e.g. staff hours, cubic meter of materials, and a fixed price per range of quantities, accommodating changes in price due to inflation or market changes.[6] This is primarily used by agencies when they have long-term projects with an estimated quantity that may be difficult to estimate, such as debris removal, levee construction, or leasing equipment for emergency responses.

There are also other qualifications to the RFP/bid request process that contracting agencies may use such as: a "set-aside" for small businesses and disadvantaged firms (women, minority, or veteran owned); or "sole-source" contracts when the need for the product is urgent, the type of providers are limited, and/or the level of product experience or familiarity with the agency's needs is limited. This latter qualification usually requires a clear written justification for the approval of a procurement authority. Lastly, some governments only procure specific types of products and services through a qualified list of contractors/vendors, i.e. Federal General Services Administration (GSA) or other government approved list, so the qualification and approval process is an additional regime that is required before they are considered for an RFP process.

When developing the scope of work for a proposal process (including an RFQ or bid request), the better practices covered in Chapter 5: Developing the Project Scope is a good starting point. In addition to this, the following are often included in the RFP process and published solicitation documentation, often referred to as a statement of work (statement vs. scope, as the former includes items below; the scope is a subset of this):

- Public advertising of the request for proposal. This may be done by posting it on the government's procurement or agency website, through industry publications or through online commercial procurement platforms. This includes a point of contact (POC) and contact information.
- Deadlines for submitting responses.
- Request to indicate interest from responding organizations, including a form with required information.
- Method for addressing questions, which may include a deadline and communication method for submitting questions. This may include a bidder conference or meeting (in-person and/or virtual).
- Required proposal format, including a required strategy, schedule, and cost estimates and budget. The latter two may require specific forms or frameworks to be included.
- List of product requirements and/or specifications.
- The evaluation method used for submitted proposals. This may include minimum qualifications from contractors and disqualifying factors.
- Experience with the type of work, previous representative projects, and the project team's professional qualifications, experience and skills related to the project.
- Type of contract to be used (more on this below), performance reporting, record keeping, and payment terms.
- Requirements for insurance and/or surety bonds, including indemnification.
- Statement of financial condition to fulfill the project scope.
- Contract agency's RFP conditions, such as the right to negotiate on contract terms with any final identified and qualified contractors or right not to conclude a contract, as well as statement of non-collusion.
- Aforementioned public agency procurement stipulations, e.g. preferences and weight for evaluation for women and minority owned businesses, "Made in America" requirements.
- The format and method for proposal delivery, whether electronic files are sent and how they are submitted, hard copies, and how many, and the mailing address and POC.

Once all responses are received by the deadline, the proposal review process begins. Most public agencies will use a cross-functional team to conduct the review and select a finalist or group of finalists. This team will often be made up of staff from the department or unit receiving the project (the client team), a member from the procurement department/unit, and any technical specialists (IT, logistics, HR, etc.). A review team will often employ an evaluation framework that includes a review matrix. This review matrix should reference aspects of the scope, quality, risk, and other project plan components as necessary. This should also include an evaluation of technical capabilities, capacity, knowledge, and experience with the product/services. Table 12.2 is an evaluation matrix template for a learning management system.

Some evaluation criteria might weight certain factors, such as quality and customer service, over costs, and will depend on the project's needs. Once the vendor is selected based on the evaluation, then the contracting process begins. Most contract types will be decided at the RFP stage and communicated with the potential contractors/vendors.

Table 12.2 Vendor Evaluation Form for a Learning Management System

Criteria Number	Category/Criterion	Plan References	Evaluation Measurement	Priority	Vendor Score (1–10)	Notes
SCHEDULE						
1.0	Time to setup and adaptation	Schedule, WBS 5	Time to set up LMS and adapt it for course delivery	6		Speak with previous customers; review during product demonstration
COST						
2.1	Software license cost	QC ID 1	Quoted cost	10		NA
2.2	Cost of hosting, maintenance, and support	WBS 5.3	Hosting + labor cost to maintain the LMS, frequency of version updates, and helpdesk support	8		Speak with previous customers on experience with maintenance and support
QUALITY						
3.1	User interface and controls	QC ID 2, Risk Plan	Rating of the user interface	4		Will evaluate based group feedback from product demo
3.2	LMS administration	QC ID 3	Rating the time needed for administration, managing user accounts, uploading and organizing courses	3		Will evaluate based on demo and feedback from other customers
3.3	Reporting function	QC ID 3	Reporting function provides user and group reports, assessments, and other self-reporting features	5		Will evaluate based on demo and feedback from other customers
OPERATIONAL SUPPORT						
4.1	Maintenance	WBS 5.3	Maintenance and can be done by internal staff	5		Speak with previous customers on experience with maintenance
4.2	Helpdesk support	WBS 5.3	Helpdesk support can be provided by internal staff. Helpdesk support required for end-users is limited	5		Speak with previous customers on experience with support
4.3	Tech. and customer support	WBS 5.3	Respond to requests within 24 h or less	4		Speak with previous customers on experience
	Total Score	**50**				

TYPE OF CONTRACTS

The type of contract used for product/service procurement will vary depending on the type of project to be undertaken. Table 12.3 identifies the contract types and what types of projects, products, or services that are suitable for each.[7]

Table 12.3 Procurement Contract Types

Contract Type	Contract	Product/Services Best Suited for Contract
Fixed Price	This contract sets a fixed price for the product/and or service. There are several types of this contract. The most common is a Firm Fixed Price, with two other variations: Fixed Price + Incentive Fee with a financial incentive for meeting schedule, cost, or technical performance requirements; and Fixed Price with Economic Price Adjustments for a multiyear contract that accommodates changes in costs due to inflation or market changes.	Most suited for projects with a well-defined and understood scope of work with limited changes expected.
Cost Reimbursable Contract	This contract pays the seller for actual costs of the products/services provided according to an agreed schedule with a fee for profit. There are variations of this contract type: Cost + Fixed Fee where the fee for performance is fixed based as a percentage of costs; Cost + Incentive Fee that includes a fee for hitting contract performance measures, and where any cost savings under budget may be shared by both the seller and buyer; and Cost + Award Fee that covers all allowable costs with an added award for meeting a set of performance criteria.	Most suited to contracts where the amount and type work may change within certain agreed parameters; with a Cost + Incentive Fee may be appropriate where an incentive is used to encourage work performance at an efficient level. Cost + Award Fee puts the emphasis on the end-product meeting a technical performance specification.
Time and Material	This contract pays the seller for actual costs of the products/services provided according to an agreed schedule, usually including a "not to exceed amount" or limit in time with contract terms that allow for an increase in budget by the contractor when needed.	This is most suited for staff or resource augmentation, or for using subject matter experts where the exact amount of work may not be clearly stated.

CONTRACT MANAGEMENT

Contract terms are usually standard, set by one party or the other to a contract, and may require negotiation on some of the terms. While contracts with vendors/suppliers are the domain of procurement specialists and legal counsel, beside the items mentioned above, some additional contractual terms contract may include as follows[8]:

- Inspection and acceptance criteria.
- Guarantees, warranties, and product support.
- Failure to perform penalties (the opposite of a performance incentives, as cited above).
- Change request process (usually by contract addenda), including flexibility to extend the contract (by cost or schedule), subject to any grant guidelines.
- Approvals for subcontractors or replacement of subcontractors or project staff.

While it depends on the relationship with the contractors, when planning the schedule, I would recommend allocating enough time at the start of the project to account for signing external contracts, i.e. subcontracts, as it will take time to write and execute contracts (sign) with legal counsel on both sides and to allocate funds for contracts. Some organizations will not allow any contractor work to begin until an agreement is in place, so any delay in contracting may result in a delay in beginning a project.

Once a contract is in place, then the contract performance and ongoing management are much the same as managing a project, with a kick-off meeting, status meetings, and tracking performance and monitoring contractual terms. For more specific guidance and practices for managing contractors, see Chapter 13: Monitoring and Control: Tracking Project Progress.

GENERAL GUIDANCE AND BETTER PRACTICES FOR MANAGING CONTRACTORS AND VENDORS

Over the course of my career, I have encountered a number of challenges when working with contractors and vendors. The following are some additional guidelines and practices that I can recommend when engaging and managing contractors:

- Ensure that the staff promised by the contract staff is performing the required work, including project management. Are the principals who are indicated in the proposal involved? Are they fully aware of all of the issues? I have hired a few contractors who included experienced and well-regarded specialists as leaders in their proposed team, only to have them disappear and leave unsupervised junior staff perform the work.
- Do not get fooled by the halo effect[9]: perception of a vendor or expert as infallible and therefore not in need of close supervision or management. While you may find some subject matter experts or contractors who have a great reputation, you still need to follow your standard operating procedures, including your procurement guidelines, and not relax your management standards.
- Make sure that you understand the contractor's general development processes (perhaps not all of the technical aspects). Questions you might pose to potential contractors:
  - What quality standards do they have in place?
  - How do they test the outcomes of what they produce?
  - How do they manage the communication with the client (you and/or the key project stakeholders, end-users, etc.)?
- The contractor should have a solid project management structure in place. Critical areas are time management (creating a realistic schedule and adhering to it), scope management, cost. The simple question I ask of vendors is for examples of their PM practices:
  - How do they manage projects? Ask them to show a sample schedule, scope, and design document for the product you are developing, if possible.
  - How do they handle clients who are unclear about their requirements?
  - How do they handle version control?
  - Also, ask for samples of the type of product you are seeking, and notice how quickly do they get back to you when asked for information or samples.
- Are they accustomed to dealing with delays in payment, especially common in governmental tax levy or grant-funded projects; can they muster working capital to get the project started? Have they planned for risks (delays, unforeseen requests, etc.).
- Make sure when selecting a vendor to get references from their last three clients (not ones they necessarily selected if possible). When performing your due diligence, see how quickly they refer these clients to you. Questions you might pose to them may include as follows:

- Are these projects they conducted for you similar to yours?
- Are they the same type of client, same industry, same size, same complexities, same content? They are bound to be different, but the more similarities the better.
- Were they satisfied with the overall process? Did the project run over the deadline, over budget originally set?
- What were the lessons learned from the project if they may share them?

- When procuring products with public-facing or employee-facing content, e.g. graphics, messaging, make sure they appear suited to the jurisdiction, whether national, regional, or local. For example, with an urban workforce there needs to be a cultural sensitivity to the end-user/target audience. This may include the needs of an urban agency working environment and worker diversity, with appropriate regional accents, terminology used, and appearance for any story-based characters, graphics, and stock photos for communication campaigns.
- Contractors should be preferred who have a deep bench (or contingencies) for all work required according to the schedule. I have encountered situations where we have waited on vendor staff to come back from vacation to complete some of the interim or final deliverables; this sometimes happened with little or no notice at the kick-off meeting or status meetings.
- If the product you are creating will need to be updated in the future, for example eLearning courses, custom software applications, is the authoring software or system proprietary? This will be critical if you decide to update or repurpose the content.
- My suggestion is that you control the design templates and graphics used and keep the level of interactivity manageable (e.g. technical design and programming requirements – see further under lessons learned below), so you can choose vendors and slowly develop the ability to manage the courses, borrowing from them or adapting as needed.
- When possible, do not give a new vendor a large amount of work, i.e. more than you can reasonably risk with a key client. Start with a smaller project and then work up from there.
- For some projects, ADA compliance may need to be met. For online webpages, Section 508 compliance may be required, e.g. video with closed captioning and audio transcript for export.

Once the contract is in place and the contractor is ready to get started, set the ground rules at the kick-off meeting as you would for a client, making sure you are both clear on the quality guidelines, expectations; make sure that you "follow them out of the gate" and they follow through to meet your expectations. Do some early checks and conduct a formal review of the early work as soon as possible to make sure that it meets your quality requirements. There may be circumstances where you may want to have the contractor collocated or in contact with the project manager on a regular basis in the early stages.

NOTES

1 Project Management Institute. (2006). Government extension to the PMBOK© Guide. p. 73.
2 FEMA's Congressional Justification for FY 2021 for Hazard Mitigation Programs: www.fema. gov/sites/default/files/documents/fy_2021_cj_final.pdf

3 Congressional Budget Justification for FEMA's Fiscal Year 2022:
   www.fema.gov/sites/default/files/2021-06/fy22_cj_fema.pdf

4 Project Management Institute. (2006). Government extension to the PMBOK© Guide. p. 71.

5 Some DHS and FEMA programs that have used an IDIQ type of contracting have been the Urban Area Security Initiative (UASI) Program, Regional Catastrophic Preparedness Grant Program (RCPGP).

6 Project Management Institute. (2006). Government extension to the PMBOK© Guide. p. 75.

7 Project Management Institute. (2017). *A Guide to the Project Management Body of Knowledge* (PMBOK guide). pp. 471–472.

8 Project Management Institute. (2017). *A Guide to the Project Management Body of Knowledge* (PMBOK guide). p. 489.

9 Refers to a cognitive bias whereby the perception of a particular trait is influenced by the perception of the former traits in a sequence of interpretations. Edward L. Thorndike was the first to support the halo effect with empirical research.

Chapter 13

Monitoring and Control

Tracking Project Progress

> Measurement is the first step that leads to control and eventually to improvement. If you can't measure something, you can't understand it. If you can't understand it, you can't control it. If you can't control it, you can't improve it.
>
> *H. James Harrington, Consultant, Quality Guru, Author*

One of the main purposes of project communications management, which we just covered, is to coordinate and track project execution and monitor performance and report on progress, keeping the project on target to achieve its goal. As the project plan moves into the project execution phase and work begins, the team and the project's vendors and contractors will begin reporting on their work progress, typically at the task or deliverable level (if subcontract) and report this to the project manager. The key areas to manage are scope, schedule, and budget (cost), as well as the other plan components, HR, risk, quality, etc., and we look at how the project is performing according to the plan.

Before we get into the specifics of measuring project performance and progress, we need to define the term baseline. A project baseline is what has been approved as part of the final project plan for execution: scope, schedule, budget, etc. The baseline allows us to measure project progress and its performance toward the end goal. Any major changes to any element of the baseline plan require a reviewed and approved change, and any deviation requires a review and corrective action, which we will get to later in this section.

EARNED VALUE AND VARIANCE ANALYSIS

How does a project manager answer the key questions: are we on schedule? Are we within budget? Are we achieving the goal? The method used is variance analysis. Variance analysis is a measure of performance by looking at the deviation from the plan, using quantifiable values. The PMBOK© cites three terms for quantifiable project value and uses basic formulas for schedule and cost variances, referred to within standard project management as Earned Value Analysis (EVA also sometime referred to as Earned Value Management). Here are the quantifiable project values:

Planned value = what is the value of the work planned (budgeted) to be completed by a certain date.

Actual cost = what was the actual cost of the work at a certain date.

DOI: 10.1201/9781003201557-15

Earned value = what is the value of the work completed to date; this is different than the budgeted amount as it states the estimated value or percent complete.

Planned value is easy to compute as it can be determined by looking at the scheduled tasks that were supposed to be completed and the planned costs (budget) as of the reporting date. This may require using a cut-off point from the schedule and budget. Actual cost is also easy to determine as that can often be derived from the organization's accounting system, looking at what was spent on the project up to the reporting date. This may require some accrual for costs spent, but not yet posted to the accounting system, in order to align with the schedule and budget.

Earned value requires a little more work. Earned value is the value based on the percentage of work completed according to the original budget as of the reporting date. As an example, if we are building concrete berms to mitigate flooding and our cost per berm is $30,000 (roughly 100 m in length at 10 ft. high), and ten berms were completed by the reporting date, then the earned value would be $300,000.

There is a challenge with computing earned value for some work products. Most common accounting systems can support actual costs vs. budget costs according to accounting line items, but do not often account for earned value, in other words a percentage complete. In addition, there are different methods for computing earned value. There are several common variations for estimating value for the work complete. If the work can be easily defined by units, then using the number of completed units as a percentage of the total can be used. If using projects tasks, then some methods use 50% for once the work has started and then 100% once it is complete; or 100% only once a specific task or deliverable has been completed. I recommend using either a unit-based method or 100% once the task is complete as the earned value, as often there is little value to an incomplete task or deliverable.

The following are the variance analysis formulas:

Schedule Variance = Earned Value – Planned Value
Cost Variance = Earned Value – Actual Cost

Let's walk through a simple example using the concrete berms for flood mitigation. As of the reporting date, we collect the following progress data:

Planned Value = $270,000
Actual Cost = $310,000
Earned Value = $300,000

This gives us the following variances:

Schedule Variance = $300,000 – $270,000 = +$30,000
Cost Variance = $300,000 – $310,000 = –$10,000

So what does this tell us about this project performance to date? Generally, positive numbers are positive performance, and negative numbers are negative performance, but it requires more analysis than that. It tells us that according to what we were supposed to complete according to the schedule as of the reporting date (Planned Value) we completed one more berm or $30,000 additional work. However, it also tells us that as of the reporting date we have overspent our budgeted cost by $10,000. This may prompt us to investigate as to why this may be the case.

In addition to direct variances according to budget, there are measures for project performance relative to the baseline schedule. This can be helpful in forecasting whether the project will end up on schedule and on budget. These formulas are as follows:

Schedule Performance Index (SPI) = EV/PV
Cost Performance Index (CPI) = EV/AC

This gives us the following project performance measures:

Schedule Performance Index (SPI) = $300,000/$270,000 = 1.11
Cost Performance Index (CPI) = $300,000/$310,000 = 0.96

Similar to the variance analysis, an index above 1 is typically positive performance and below 1 is negative performance. A schedule performance index of 1.11 would indicate that the project is performing roughly 11% ahead of schedule. The cost performance index of 0.96 would indicate that the project is approximately 4% over budget. This would be at the time of reporting, not necessarily for the entire project. Again, as with any performance analysis, we need to evaluate the root causes that may be impacting performance further.

As we mentioned above, we may want to forecast whether the project will end up on schedule and on budget based on current performance. This is a little more complicated than it may seem, as there are several formulas for computing this. The formulas for calculating whether the project will end up within budget are given below. Note – for Budget at Completion (BAC) is the planned budget for the project. For this example given in Table 13.1, the budget at completion is $1 million.

The first two formulas weight the actual costs and cost performance, so provide a more pessimistic forecast for the budget. If we figure that the project will completely deviate from our plan or performance to date, then we can redo the estimates for any work that

Table 13.1 EVA Estimate at Completion Formulas

Forecasting Use Case	Formula	Sample Forecast
If project performance is expected to continue according to the CPI	EAC = BAC/CPI	EAC = $1 mill./ .96 = $1.04 mill.
If project performance is expected to continue according to the planned rate	EAC = AC + BAC – EV	EAC = $310,000 + $1 mill. – $300,000 = $1.01 mill.
If project performance is expected to deviate from the plan or current performance	EAC = AC + Bottom Up Estimate to Complete (ETC)	**Can't compute without additional estimates.**
If project performance is expected to continue according to both the CPI and SPI	EAC = AC+([BAC – EV]/(CPI × SPI)]	EAC= $310,000 + [($1 mill. – $300,000)/1.11*0.96] = $966,907

Source: Project Management Institute. (2017). *A Guide to the Project Management Body of Knowledge* (PMBOK guide). p. 267.

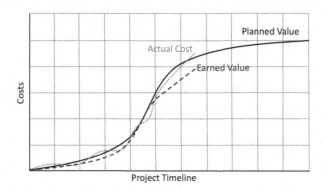

Figure 13.1 Earned value management – Work performance analysis.

needs to be completed. The last formula weights these two indices equally, so the forecast is more optimistic.

Data points at a single point in time only provide a snapshot of project performance. Trend analysis can be more insightful in helping determine project performance using a trend line of multiple data points. The example in Figure 13.1 provides a trendline for the three EVA performance measures: planned value (budget), actual cost, and earned value. This provides us with an aggregate measure of how the project is performing over time. Looking at the example, we may determine that over time actual costs have stayed close to the planned value, so while we might monitor actual costs (in other words cost variances), we may be less concerned with them. However, earned value appears to be lagging behind the planned value, so we may want to investigate what may be causing this difference and determine whether the project will wind up on schedule and on budget.

Ultimately, forecasts are only as accurate as the data we collect, and what is just important is to probe the factors that are influencing project performance and what corrective actions or changes to the baseline that we may need to make. When projects deviate more than the risk plan allows, the project manager may need to escalate this to a more senior level, for example to the project sponsor and client level.

When it comes to project cost and schedule forecasting, one area of research to be aware of is the 5% rule.[1] This rule states that once the first 5% of the project is complete, the scope and risks are better understood, and the remaining 95% of the project can be estimated with greater accuracy. The 5% rule has been shown to be accurate in forecasting the project end date and budget according to a number of studies. The 5% rule can be helpful for managing projects in their early stages, as it supports greater attention to identifying and mitigating risks early in the project lifecycle to ensure it meets its targets.

NUTS AND BOLTS OF PROJECT TRACKING

Project performance reports and analysis help to inform project status and review meetings. Depending on their preference, the project manager will ask each team member to report out on their current task status and also present how the project is progressing. A template with a sample report on how a project manager would record progress from a status meeting is given in Table 13.2, again, using the software application project.

Table 13.2 Project Tracking and Corrective Action Plan

Project name	Web-based Software Application Project for a Special Medical Needs Registration
Meeting participants	Clarissa Chang, Project Manager; Allison Ford, Lead Software Developer; Fred Murakami, Quality Manager; David Greenberg, Application Architect; Sandra Kumar, Application Developer; Katherine Yen, Database Developer; Dmitry Vucevic, Web Developer; Kevin Jackson, Software Quality Assurance Specialist
Date/location	1/31/2X at ABC Co., 42 Main St., Room 302A

Project Progress

Project Plan Component	Status (On Track, Off Track; if so, Variance)	Action Required – Y/N?
Project scope	On track, no change requests	N
Project schedule	Ahead of schedule, schedule variance +2% from the baseline	N
Project budget	Off track, budget variance –2% from the baseline	N
Quality	QA testing under development; first round of testing scheduled	N
Risk	LMS cybersecurity risk raised by Liberty City IT Director	Y
HR	All team members hired and first round of training complete	N
Procurement	Learning development contract and LMS contract in place; up-front payment for contracted staff	N
Stakeholder/ communications	On track	N

Issues

Issue Description	Project Impact	Action Required?
LMS cybersecurity risk	Risk was not previously identified.	Yes
Upfront payment for contract	Up-front payment for contracted staff member (web developer) was not anticipated so early and increased costs slightly, but offset by work completed sooner.	No

Action Items

Action/Decisions	Issue/Plan Status Related	Responsible	Target Date
Meet with LMS provider for eLearning course to analyze cybersecurity risk	Risk plan	CISO, Bruce Todd	2/10/2×
Review budget and costs for contracted staff and work performance to project final costs for tasks	Budget	Project manager	2/5/2×

A common approach to monitor progress and manage work is called "Management by Exception". This is addressing any items that are significant plan deviations or changes to the existing plans, newly surfaced risks or quality requirements. Any tasks that are proceeding according to plan are left alone, but any work that has deviated significantly and warrants attention is analyzed and addressed with a corrective action as needed. The final step in this process is to document, implement, and manage the corrective action in order to keep the project on track according to the original plan.

PROJECT INTEGRATION MANAGEMENT AND CHANGE CONTROL

For any deviation from the plan or scope change, we need to make sure that we document it and to review how that action may need to integrate with the other elements of the project plan. Ensure that change control measures are understood by the team and key stakeholders, project sponsor, and client team. While it may seem trifling, if the scope is changed, even in a small way, then it needs to be reviewed in terms of the other plans (quality, cost, HR, procurement) and change in contract or scope applied. Referencing the integration of the WBS to the other parts of the project plan, we can see in Figure 5.3 (Chapter 5) that if we make any changes to one part, then it may impact another.

For example, you are managing a project to develop job aids and training for an emergency plan, and an agency client has their legal department review the emergency plan late in the development cycle. The legal department wants to include a proviso in the plan document that they make no guarantees that the resources or services will be provided by the government to its residents. Here is the cascade of changes:

1. This impacts the scope of work, as the job aids and the training on which it is based will need to change.
2. It also impacts the procurement plan as several subcontractors now need to change the design of the job aids to accommodate this addition; the same with the training courses.
3. It also requires a review and update to the risk plan, to see if this was a risk was planned for, and the same goes for the quality plan.
4. There may be an adjustment to the schedule and budget as this may require additional work.
5. We may want to look at the communications plan to ensure that the legal department has been accounted for and to adjust their level of influence, when and how they are managed as a stakeholder, the frequency and messaging for communication.
6. Lastly, we may want to look at the HR plan to see how this may impact our team.

In sum, any change that impacts the baseline plan needs to be reviewed carefully according to the entire plan to ensure that the change can be accommodated and, if accepted by the client/sponsor, incorporated into a new and updated baseline plan.

MANAGING CHANGE CONTROL WITH PROJECT FILES

We covered how project files might be organized in the previous chapter on communications. Now we need to turn to document management, as seemingly simple as that may seem. During my career there have been some instances where project document version control became an issue due to the following:

- reviewers were working with the incorrect version of a document;
- there was confusion as to what version was the latest; or
- changes were not recorded properly, written over, or erased by subsequent reviewers.

It is a frightening situation to arrive at a work computer in the morning to discover that a long document with exhaustive reviews over the period of weeks from a client team is missing their edits due to the wrong document version.

While file-sharing systems often have features for handling edits from multiple individuals, you still need a set of rules and procedures in place in order to avoid version control problems. Here are some simple guidelines and procedures for managing documents:

- Make sure to confirm who will be reviewing the document at the start of the development document review process, in particular from the client team. This should be made clear in the stakeholder and communications plans but keep an eye out for guest reviewers who were not included during the requirements definition phase and in the review plan. Guest reviewers can cause delays due to the need to re-explain the project and possibly accommodate additional or different product requirements.
- Use a development stage in some part of the electronic file name for product development documents (including any zipped files) and give a date at the end. For example, **Alpha-COOP Plan_202X.01.31**; include a date at the end in the form of YEAR-MONTH-DAY, so sorting versions from more recent to later versions is easier.
- If using a document-sharing program, all edits and comments are typically tracked according to the individual user, but it helps if you are new to a system to check it at least once at the beginning to make sure that the tracking feature is working properly.
- If you are not using a file-sharing system, then you want to carefully track edits from different reviewers. One method to do this is to ensure that once an individual has reviewed the document, they add their initials after the file name they save it under, e.g. **Alpha-COOP Plan_202X.01.31-ADB**. Additional reviewers add their initials after the first reviewer, e.g. **Alpha-COOP Plan_202X.01.31-ADB-CSK**; this way you know who has reviewed the document.
- Make sure the end of the review period is clear for the reviewers and include that message in the communication when sending it out for review, for example "Please ensure that all reviews are completed and final by COB (Close of Business) 3/21/2X".
- Make it clear how you want edits to be made, whether using "tracked changes" (for MS Word documents), using notes, or other features of the shared document review system. Ask for feedback that is specific and actionable. Also, make it clear how edits will be made and how you want any conflicting feedback between reviewers to be deconflicted, whether by the client review team, the recipient, or by group review. As a project manager, I usually ask the POC from the client team to deconflict and consolidate all comments and edits before they are submitted back to our development.
- When the edited document has been received, the document manager on the development team will want to review the edits and comments as soon as possible and make sure they are clear enough to edit. This may require reaching back out to the individual reviewer to clarify any unclear or confusing feedback. Once the revisions are made, then you want to make sure to save the new document under a new development phase and/ or date as appropriate and begin working on those edits for the next version.

ORGANIZED PRODUCT REVIEWS: WALK-THROUGH, PILOT/SYSTEM TESTING, AND TRIAL RUNS

Some product reviews require more formal organization and facilitation in order to elicit and manage constructive feedback. Product versions and types that constitute multiple deliverables or require user instruction and/or an interface such as equipment, technology and software applications, and eLearning courses can benefit from feedback from product end-users. A general rule in conducting end-user testing, trials, or pilots etc. is to ensure that the end-users selected are representative of the audience who will eventually use the product.

A walk-through is a facilitated review of the product, facilitated by the Lead Developer of Project Manager. It may involve sending the product out ahead of time, for example plans, job aids such as field operations guides, or course materials or online programs. A walk-through functions like a guided tour, pointing out functions and features that related to the product quality requirements. A product walk-through often precedes a pilot or system test.

A pilot test or system test is usually used for courses, both in-person or eLearning (online), software applications, some equipment, and, occasionally, job aids. A pilot or system test requires scheduling a location, suitable system configuration such as computers and network or internet connections and access to the application/eLearning (hosted on an LMS). There are different types of software application testing (unit testing, black box testing, UX testing, etc.), and the conditions and methods used for each are particular. Since this book is not software focused, we will not go into depth on this type of software testing.

A trial run is similar to a pilot, except it may sometimes be conducted under operational conditions out in the field or in a similar environment, with some controls or constraints put in place. It may be most suitable for testing equipment or incorporating new technology, such as the use of UAVs (drones) or GIS capabilities. What is key is setting up the trial testing conditions as close to how it would be used in an actual emergency or public safety response environment. In some development methodologies, a pilot/system test, conducted in a controlled environment with a limited number of users, may precede a trial run, conducted in a less controlled field or operational environment which may subject the product to more strenuous conditions. This is done as the development stages align with the appropriate testing environment, not pushing the version beyond its limits before it is ready for the field. In some cases, there may be multiple trail-runs before the product is suitable for launch.

There are several guidelines and methods that help to frame and obtain optimal feedback from organized product reviews:

- A set of written guidelines explaining the development stage and the intent of the type of review. This may include a note describing the role of the end-user involved in the review or test and what they can expect, how much time may be needed to participate, and how their feedback will be handled.
- An evaluation form may be employed that includes a set of quantitative and qualitative feedback components.
- Some reviews may also include an organized oral debrief with end-user participants. If it can be arranged, I recommend bringing in someone outside the development team to help facilitate this type of end-user participant debrief to help avoid bias from any

development team member in how they might elicit feedback; it also helps the end-users feel that they can be more honest with the facilitator as they do not have a stake in the product.

● Facilitation of any review needs to engage the end-users to get their full participation and feedback. Make sure to schedule it on a day and time that maximizes their energy and attention, avoiding holidays, weekends, and the possibility of running late in the day, especially with governmental agency staff who may be used to working a standard work day.

There may be situations which require a formal sign-off, either a part of the contract terms or necessary to confirm the completion of a client review to prevent a runaway review cycle. This can be handled manually with a scanned hard copy or using an electronic signature, although I recommend consulting with an organization's legal counsel to ensure that the method used holds legal weight. While seemingly basic, putting a clear set of rules in place for document and development versions reviews can help make a project manager and their team's work easier.

PROJECT CLOSURE, TRANSITION, AND INTEGRATION INTO OPERATIONS

There are still a few more steps to go before your team has submitted the final project deliverables, and the client team has accepted and signed off on them as complete. We began this book discussing the origins of projects, which are ultimately to build the capacity of an emergency or public safety agency in their ability to prevent, mitigate, and manage incidents. A project is only truly complete once the results have been fully integrated and made operational in an organization. This transition into operations is part of the project closure process. Even if a project is stopped early, project close out is still part of the concluding a project.

Although they may differ, depending on the type of project and an organization's procedures for project closure, here are some of the major steps in project closure:

1. A handover to the client team of all product documentation, including designs (blueprints, design documents, system configuration, technical specifications, etc.), instruction manuals, root files (for graphic files, copy-ready e-files for artwork, online applications, etc.). As some of the files may be large, sharing them through the use of shared folders is most common. Some contracts stipulate a specific set of hard copies of final documents, such as plans, job aids, and manuals.

2. Ensuring that the staff who will assume responsibility for the new capability (equipment, system, plan, etc.) have incorporated it into their standard operating procedures (SOPs). This includes adding technical support for any new systems and technology, procuring any stock of spare parts and consumables, and scheduling any routine maintenance and updates. This latter area is especially important for software projects which may require updates with new operating systems, network configurations, and security patches.

3. Closing out any procurement contracts, including payments to vendors/contractors, and releasing any contracted staff, along with any final performance reviews.

4. Updating all project documentation, including confirming that the scope of work has been fulfilled, any final risks encountered, quality issues addressed, as well as any HR changes or adjustments (team member replacements or early release of team members). For the lessons learned/AAR (see below) and any future review of the project, including any audits, it is critical to properly archive the project records, including backup folders.

5. If funded under a governmental grant, confirming the disposition of any equipment purchased through the project. When it comes to computers equipment, which is typically amortized over a period of one year or slightly longer, most donor organizations cannot absorb additional equipment and do not want to have it returned, so they will often donate it to the recipient organization if the performing organization is a non-profit organization (NGO, educational institution, etc.). For any equipment or systems that can be amortized over a number of years, a disposition plan should be developed and submitted to the client organization for consideration by the donor organization for their final decision on its disposition.

6. Dispensing with any excess products, materials, or supplies. As with computer equipment, the value of these items is often not worth the effort for the client organization, but items such as additional hard copies of job aids, instructional manuals, or plans may be valuable to the client organization.

7. Conduct reviews of team member performance, acknowledge their contribution to the project, and release team members from the project.

8. Conduct an After-Action Project Review, often referred to as a Project Lessons Learned Meeting, to identify areas of improvement, confirm and enhance existing project methodology, and to adopt new practices to address any performance-related challenges experienced during the project. We will cover this in more detail in Chapter 16: Quality Management in Emergency Management Programs and Continuous Improvement for Responses, Projects and Programs.

9. Confirming that all final project contract payments have been made to the performing organization.

10. Closing out the project contract with the client POC, project sponsor and both internal legal counsel and the client organization legal counsel to acknowledge all contract requirements have been met.

Much like bringing a ship into dock, we want to bring the project to a smooth conclusion, making sure that all contractual obligations have been fulfilled and documentation archived. We also want to acknowledge a job well done and any areas of improvement, as conducted through a project lessons-learned process. While it may appear to be mundane, administrative work, all of this work to close out a project helps to institutionalize project management methods and formally end a project.

NOTE

1 Clements, D. P. (March 2004). The 5% Rule: A Simple, Accurate Way to Estimate Project Completion *Journal of Construction Engineering and Management, Volume 130*, Issue 2, Pages 134–143.

Part 3

Strategic Project Management

Chapter 14

Program Management and Project Portfolio Management

The LAFD is taking the initiative to develop ideas that will make the fire department more sustainable and environmentally friendly. New construction of fire stations incorporate energy saving features. Opportunities and initiatives are sought to encourage water conservation, Such as discontinuing lawn sprinklers and recycling water during the engineering examinations. Community resilience is the ability to prepare for anticipated hazards, adapt to changing conditions, and withstand and recover rapidly from disruptions. Activities, such as disaster preparedness—which includes prevention, protection, mitigation, response and recovery—are key steps to resilience. LAFD programs such as community risk reduction and Community Emergency Response Team (CERT) help residents proactively protect themselves against hazards, build self-sufficiency and become more sustainable.

Fred J. Mathis, Chief Deputy, Administrative Operations
Los Angeles Fire Department, LAFD a Safer City, Strategic Plan 2018–2020, p.28.

There are two approaches when managing projects collectively: program management and portfolio management. When projects have aspects that relate to one another, such as design elements, end-products, or long-term support services that would benefit from being managed collectively, this presents an opportunity to manage them as a program. As defined by the Project Management Body of Knowledge, a program is a group of "related projects, subsidiary programs, and program activities that are managed in a coordinated manner to obtain benefits not available from managing them individually".[1] A program can also be a mix of projects and ongoing operations that relate to the projects undertaken. A program can be part of a portfolio.

As Figure 14.1 illustrates, both project portfolio and program management are intended to plan and manage projects in alignment with the organizational strategy and objectives.

Project portfolio management is the centralized management of a portfolio of programs and projects to achieve organizational objectives.[2] Similar to financial portfolio investment that seeks to increase value through a "Return on Investment" (ROI), project portfolio management is employed to efficiently manage funding from grants and tax levies for programs and projects that are planned in order to return value to the community from the collective benefits achieved from the portfolio. A portfolio contains programs and projects that may or may not be related to one another, so portfolio management practices seek to optimize resource management within available capacities and capabilities, while balancing the risks of achieving those benefits, to return value to the community.

DOI: 10.1201/9781003201557-17

Figure 14.1 Strategic, operational, and project planning.

Note, as with other management practices, both portfolio and program management have their own standards and build on the principles and share some of the practices of project management. This chapter is intended to cover both portfolio and program management as they relate to emergency and public safety projects. They are presented in this chapter at a level at which a project manager can manage or integrate the project into the overall program and portfolio management process. It is not intended to comprehensively cover all aspects of portfolio and program management methods and practices.

PROGRAM MANAGEMENT AND THE PTE CYCLE

One of the best examples of a program related to emergency management is the Planning, Training, and Exercise (often referred to as PTE) cycle, as seen in Figure 14.2.

Each emergency management and public safety unit, department, or agency has an ongoing cycle of preparedness that resembles this cycle, whether it is done informally or formally planned out. We develop emergency operations plans at the start which then requires staff training, followed by exercising the plan and training, and results in an evaluation (after-action review and report) which feeds the improvement plan (note – the AAR/IP is often a collective document).

An example is given in Figure 14.3 that illustrates how a public health or emergency response agency might develop a strategic response to a new pandemic. The starting point is utilizing lessons learned from an AAR/IP conducted after the emergency response to the COVID pandemic. This is displayed vertically from left to right (as a hierarchy) for ease of viewing. Let's walk through the general structure:

1. At the very top, there is a Strategic Plan for Enhancing an Epidemic or Pandemic Response. This defines the strategic objectives of how the organization will respond to an infectious disease outbreak.
2. This is broken down into four key project areas that relate to the preparedness cycle:

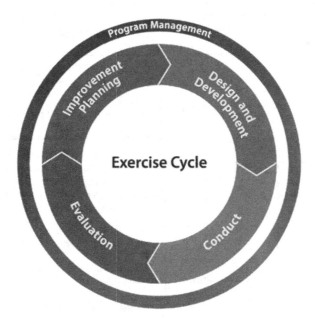

Figure 14.2 Emergency program: planning–training–exercise cycle.

- Updating Pandemic Response Plans.
- Procuring and Organizing Resources.
- Revising Training.
- Developing a Pandemic Exercise Program.
3. Each of these is broken down into subprojects.

Now, let's review some of the areas where benefits might be obtained from coordinating these projects, as opposed to managing them independently.

- When updating the epi-data reporting procedures and communications, there may be data obtained from the field, in particular testing and vaccination data from the POD plan (as part of a mass prophylaxis plan to an epidemic or pandemic), where reporting procedures within those respective plans would benefit the epi-data reporting.
- There may be common procedures planned sites such as Pods, testing sites, or isolation and quarantine, for example requirements for safety and security, DAFN (ADA) compliance, and common logistics, for example flows for car and pedestrian traffic, and supply management.
- There may be common resource needs between different response plans, e.g. traffic cones for vehicle traffic at sites, vests for staff, and tables and tents for cover outside, so consolidating the procurement and contracting for those needs in a coordinated manner provides efficiencies for the organization. Moreover, there are savings for volume purchases and standardizing equipment and supplies for multiple types of responses (functions that meet an all-hazards approach).

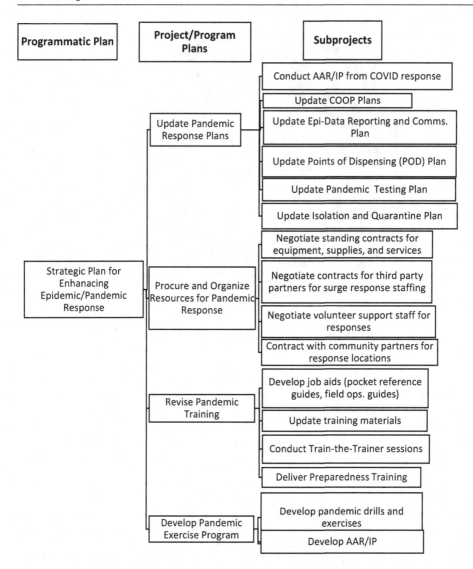

Figure 14.3 Sample strategic plan for epidemic/pandemic program response.

- There are benefits gained from developing and using a common evaluation framework for public sites used for PODs and testing (perhaps other responses), as well as using staff for site evaluations.
- Common designs for job aids (pocket reference and field operations guides for SOPs) may be used between different response plans. This makes it easier for cross-training staff on different plans as the user experience will be similar, so finding sections and following the formatted directions is easier. This is also true for developing training materials, Train-the-Trainer (TTT) sessions.
- Drills and exercises can leverage outlines and content from a common core, e.g. exercise documentation such as an ExPlan, MSEL, and EEGs can be repurposed from one another.

THE BASICS OF PROGRAM MANAGEMENT

So, what does planning look like when it comes to coordinating these various projects, subprojects, and subprograms as one program? The Project Management Institute (PMI) publishes the Program Management Standard[3] which contains better practices in the area of program management. There are a number of similarities between project and program management, such as scope, schedule, and cost management. However, programs differ in several key areas:

• Programs are typically a complex grouping of interrelated subprograms, projects, and subprojects performed over a number of years (or decades for large engineering projects) to deliver benefits (value) to the organization.
• Program benefits collectively align with organizational strategic objectives and those of the community that the program serves, and the benefits may be described by both measurable, financial benefits, e.g. reduced costs from natural hazard damages; or benefits may be difficult to measure, e.g. reducing harm to life-safety from natural hazards. Hence, a program's "Cost/Benefit", sometimes referred to as the "Return on Investment", may be difficult to measure in tangible benefits or may have a return in a more distant future, over many years or decades before it is realized.
• With their complexity, programs have a greater demand for integration and coordination, requiring increased communication and time for planning between projects in order to consolidate requirements, synchronize designs and work, and optimize the expected benefits. This includes coordination and collaboration with stakeholders, especially for public-facing projects where this may include community hearings and presentations, budget approvals, and risk assessments, including environmental impact studies.
• Programs may have a greater degree of change management due to their longer lifecycle and, in many cases, different streams of grant funding and reporting requirements. Changes in political bodies at local, tribal, territorial, state, and Federal/National levels, can influence grant funding and/or requirements, which can influence the program and its components. There is also the ever-present possibility of a disaster impacting programs in a variety of ways.
• Programs require sustaining over a longer time horizon, maintenance, and support to keep them current and operational. This requires planning throughout the program lifecycle, in particular the latter part of the program. Budgeting for activities that will integrate and operationalize new capabilities is especially important to sustain program benefits.

Figure 14.4 illustrates the program management lifecycle.

Let's also take into consideration that some of these elements may have different funding sources, may be managed by different project managers, or fall under different departments, and have different stakeholders, clients, etc. However, they may share a common governance structure (project supervision), and common end value they are seeking to achieve, and a common set of resources as described above. Using common resources or engaging similar stakeholders may result in potential bottlenecks when it comes to developing and

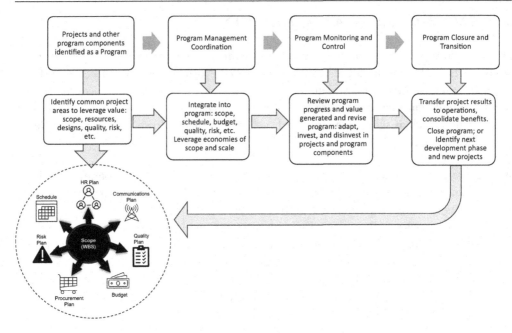

Figure 14.4 Program lifecycle value management.

reviewing project work, which offers another benefit for managing these as a coordinated program.

Another good example of relating the program management concepts to a case is the Regional Catastrophic Planning Team (RCPT) of NY-NJ-CT-PA that was funded under the Regional Catastrophic Planning Grant Program (RCPGP) from 2008 to 2015. The RCPT was a multi-jurisdictional group that helped to guide and manage the RCPGP-funded projects and programs. The RCPT NY-NJ-CT-PA multi-jurisdictional group included[4]:

- Northern New Jersey Urban Area Working Group
- New York State Division of Homeland Security and Emergency Services
- New Jersey Office of Homeland Security and Preparedness
- New York State Office of Emergency Management
- New Jersey State Police/Office of Emergency Management
- New York State Office of Counter Terrorism
- Connecticut Department of Emergency Services and Public Protection
- Nassau County (NY) Office of Emergency Management
- New York City Office of Emergency Management/New York City Urban Area Working
- Group
- Suffolk County (NY) Fire Rescue and Emergency Services
- Westchester County (NY) Office of Emergency Management
- Pike County (PA) Emergency Management Agency
- Port Authority of New York and New Jersey Office of Emergency Management
- Citizen Corps Council and Metropolitan Medical Response System
- Private Sector

There were other RCPGP-grant-funded sites in major metro areas, such as the Bay Area and Chicago Area. Each of these grant-funded sites focused their catastrophic planning and preparedness efforts on different areas, unique to their own needs. The RCPT NY-NJ-CT-PA focused its catastrophic preparedness work on "fixing shortcomings in existing plans, building regional planning processes and planning communities, and linking operational needs identified in plans to resource allocation".[5] This RCPGP-funded NY-NJ metro region site was the largest in the country, with 30 different counties in three states, with three different FEMA regions, one of the many complexities in planning these efforts under one umbrella. One of the major planning considerations for a catastrophic event in this RCPT site is that jurisdictions would be competing for some of the same resources in responding to and after a catastrophic event. The RCPT was organized in a manner that helped to guide and manage the overall, multi-year program.

Figure 14.5 provides a simplified, notional program organization for the RCPT. Note – this is to illustrate the general organizational structure, so not all program or project areas covered under the RCPT program have been listed.

The steering committee constituted representatives from the different jurisdictions that made up the RCPT. Reporting to the steering committee was the Regional Integration Center (RIC), which was the operational arm of the RCPT, helping to define and manage the specific programs and projects that would meet the grant program's strategic objectives. There were other layers of program and project management that supported this work, such as plan advisors and plan leads, who were part of the development teams to collaborate and provide feedback to the contractors hired to develop these programs and projects.

Among these strategic programs and projects were[6]: .

- Regional plans, playbooks, tools, and training
  - **Critical infrastructure resiliency:** Supports the restoration of electric power after a major outage, including an Executive Level Power Outage Primer.
  - **Debris management:** Focuses on the steps of debris management and response and includes a guide and DOZER stand-alone software application designed for debris management.
  - **Evacuation coordination and emergency sheltering:** Supports the evacuation decision-making process and framework for regional coordination, and evacuation best practices, emergency shelter needs, specifically for children's needs, disabilities, access, and functional needs, service animals, and household pets, including a web-based, emergency shelter management system.
  - **Housing recovery:** Provides a playbook and Housing Recovery and Rapid Repair program (H3R) to rebuild damaged housing following a catastrophic event.
  - **Mass fatality:** Helps jurisdictions coordinate mass fatality management (MFM) and includes an MFM Catastrophic Assessment Team, MFM Field Operations Guides, Unified Victim Identification System (UVIS), and MFM response system training.

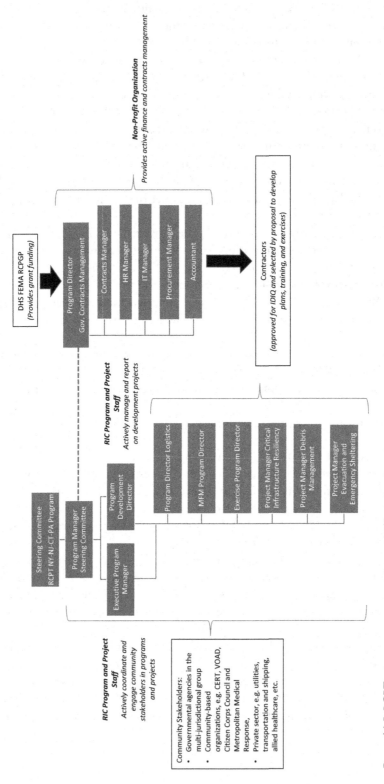

Figure 14.5 RCPT program management organization structure.

- **Nuclear incident (IND/RDD) response:** Provides a framework and planning tools for IND response, including defining a Science and Technology Advisory Committee to address guidance and communication.
- **Private sector integration:** Addresses the needs of private sector requirements, establishing a Business Coordination Center and Business Coordination Center.
- **Volunteer coordination:** Creates a mechanism to coordinate with volunteer organizations, matching requests with volunteer organizational resources.
- **Regional logistics program:** Develops and promotes the implementation of a universal standard in disaster logistics, including plans, field operations guides, intermodal analysis, and training.
- **Access and functional needs (AFN):** Provides planning considerations and guidance to assist with accommodations for people with disabilities, access, and functional needs for all emergency facilities and services.
- **Continuity of operations plans (COOP):** COOP plan for a single site to regional COOP response and templates to solicit proposals for a COOP Plan.
- **Planning for regional coordination and EOC support:** Develops Federal Integration Plans (FIP), Joint Field Offices (JFO), the concepts of Unified Area Coordination Group (UACG) and Forward Operating Base (FOB), ESF job aids, and a dashboard for situational awareness to EOC executive decision-makers. This includes EOC training.
- **Exercise program:** This includes an Emergency Management Catastrophic Exercise Program and Lightning Bolt Exercise Program, providing turn-key exercise capabilities to evaluate regional coordination and response at strategic, EOC, and field response levels.

What does the planning for all of these program components look like? In addition to the practice areas that we have covered under project management in the previous chapters, the following are additional, basic program management planning practices and some examples of how they might be planned for through the RCPT program:

- **Identifying and analyzing stakeholders:** In order to define the program benefits, program stakeholders need to be defined and engaged. This included the various governmental partners and their constituent agencies and organizations. Concurrent with high-level strategic planning to define overall program benefits, the RCPT held meetings with these stakeholders to elicit and incorporate their program requirements in order to elaborate and integrate these planned benefits into the program.
- **Identifying and analyzing benefits expected from the program and linking to components:** One of the first steps in developing a program is to identify and analyze the expected benefits tied to the program's business case. The benefits identified may include both quantifiable outcomes, as well as qualitative outcomes that are more difficult to measure. For the RCPT, some of the key benefits for the regional community would be:
  - Increasing the capacity for regional emergency response agencies (federal, state, and local) and community stakeholders to respond to and manage the

cross-jurisdictional and acute impacts of catastrophic incidents. This could be measured in qualified staff trained on new plans, new systems, and equipment and supplies.

- Integration with and interoperability between regional emergency response agencies and community partners. Over 700 participants across all levels of government, the private sector, and community-based organizations were engaged in the development of the regional plans.[7] A good example of this was the development of the Universal Logistics Standard, with the goal of building a comprehensive disaster logistics program for the region.

- **Coordination with budgets, obtaining grant and tax levy funding:** Program budgeting is inherently more complicated and requires a greater degree of long-term planning, coordination, and management. FEMA grant funds are always transferred to a state agency and then to local organizations. The planning arm of the RCPT, the RIC, managed funding from the US Dept. of Homeland FEMA. In the case of the RCPT, contracts and funds were directly managed by a national non-profit organization, with supervision and approvals by the RIC. A standard FEMA grant requirement, cost sharing required for RCPGP grant funding, was met through time contributed to the RIC staffing and work from the emergency management agencies involved (from their tax levy paid employees from the multi-jurisdictional group). The RIC in concert with the non-profit agency coordinated disbursements to hired contractors/vendors for the work and managed reporting to DHS FEMA.

- **Coordination of program component schedules, teams, contractors/vendors, and partners:** The RIC Staff took an active role in managing contractors/vendors, assigning specific staff to manage individual projects, and coordinating work with contractors and jurisdictional agencies. Coordination required intensive logistics planning for in-person training and exercises. The RCPT continually held briefings and coordination meetings with key regional stakeholders, the multi-jurisdictional group, community-based organizations, and the private sector, such as major utilities and commercial companies.

- **Change management:** Over the course of the seven years of active program planning and development, a number of changes modified the original strategic program components. Two major emergency incidents took place during the program tenure, Hurricane Irene (2011) and Hurricane Sandy (2012). These two major disaster events brought greater consideration to the needs of people with disabilities, amplifying attention to projects focused on building capacity to accommodate AFN. In addition, these two events highlighted the need to fill gaps in housing recovery for the region.

A notional example is given in Figure 14.6 for the RCPT program; this is not an actual schedule for these program components but a representative schedule of this multi-year program.

At the conclusion of the program, interagency coordination structures were established to continue maintenance and support for program outcomes and their benefits.

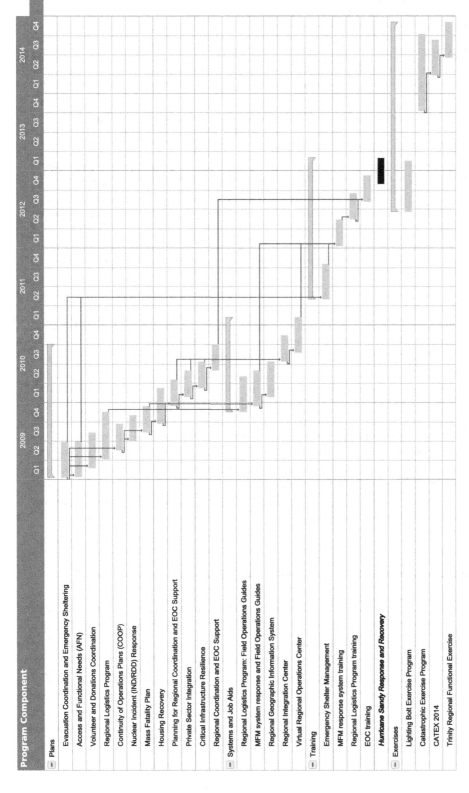

Figure 14.6 RCPT Program Schedule (Notional).

Cybersecurity Efforts in the San Francisco Bay Area

The Bay Area Urban Areas Security Initiative (BAUASI) region comprises over 650 municipalities with a population of 7.7 million residents in the greater San Francisco Bay area.[8] Many of the leading global internet companies have their headquarters or have offices in the "silicon valley" of the Bay Area, so safeguarding critical infrastructure sites from cyberattacks is a top priority.

The BAUASI grant funds the Northern California Regional Intelligence Center (NCRIC) to enhance capabilities in protecting the region's public cyber infrastructure through constant collaboration in sharing cyber threat intelligence, improving understanding of cyber-attacks, and developing models to counter them.[9] The Cybersecurity Work Group works with the NCRIC to coordinate with Chief Information Officers, Chief Technology Officers, Chief Information Security Officers, and Information Systems Managers from each of the UASI jurisdictions. Their objective is to develop regional capabilities in detecting malicious cyber activity by sharing best practices to increase cyber resilience and conducting technical countermeasures against existing and emerging cyber-based threats, facilitating quicker recovery for victims of cyberattacks.

In 2020, the BAUASI Cyber Resilience Work Group and the NCRIC Cyber Team jointly developed the Cyber Incident Response Framework to address both OES and IT objectives, supporting county-wide cyber incident response efforts. Additionally, they produced a Cyber Toolkit comprising tools and templates to assist individual organizations and jurisdictions in strengthening their cyber planning, response, and recovery.

The NCRIC played a crucial role in ensuring the security of the 2020 election cycle by conducting 400 cyber briefings for regional partners to explain potential threats. Although there were several cyberattacks from foreign actors in the days leading up to the election, the election process itself remained unaffected, and according to government reports, the 2020 election was the most secure in history. Furthermore, the Cyber Team extended critical cyber incident support to 15 ransomware attacks in the region. This support included information triage, building cases for federal partners, and sharing threat information with stakeholders to protect potential victims.

ENGINEERING PROJECTS FOR MITIGATION AND REBUILDING

Large-scale engineering projects involve deep and broad levels of technical expertise for mitigation, whether to prevent flooding, decrease exposure to wildfires, or reduce impacts from earthquakes; or for rebuilding during the recovery phase post disaster. Hazard mitigation and rebuilding projects often require extensive and lengthy environmental impact studies and analyses, public hearings, and regulatory filings before they may proceed.

These types of large-scale engineering projects differ from some of the examples presented so far and are exclusively undertaken by large firms with extensive experience, knowledge, and a well of resources to draw upon. Some of the key capabilities required by these firms to develop and execute these projects are:

- Subject matter expertise in risk assessment for hazard- and threat-related areas, for example in the science of fire, flood, and wind effects, and the potential for damage and measures for prevention and mitigation. This may include proprietary or access to commercial computer modeling for these hazards.

- Administrative capacity for handling the paperwork required to complete the RFP or bid process. The public bid and RFP process can be quite burdensome, and firms competing at this level can expect to expend many staff hours to submit a bid proposal, often without compensation if they lose, so they need to have deep pockets to cover these unreimbursed costs, knowing that they will win only a small percentage of bids or RFPs but certainly not all of them. In addition, they are usually required to have sufficient insurance and bonding (performance-related) coverages.

- Firms that compete on public bid requests act as the general or prime contractors and rely on a well-developed project management talent pool and staffing resources (internal and contract staff), with extensively sophisticated supplier and subcontractor networks to meet the needs of these types of specialized projects.

- By their nature, large engineering firms are organized on a project basis, with finance, HR, and other functions serving as support, and not competing for resources or the attention of senior company leadership. This ensures a higher level of visibility for project managers in the organization, as well as accountability and support. Indeed, senior leadership is most often drawn from the ranks of successful, experienced project managers in the industry[10].

- The systems, including both managerial and information technology, are sophisticated and adept at managing project finance and scheduling.

- Institutional knowledge and professional practice are highly valued and maintained, especially as they influence the organization's industry reputation.

After a major disaster has occurred and the wreckage and debris have been removed, community recovery through rebuilding involves a complex web of damages assessments, insurance claims (both private and public, e.g. the National Flood Insurance Program), government aid, building permit approvals, and contracting for reconstruction. The current mantra of "Building it Back Better" is commonly the call during the long-term recovery phase, especially following catastrophic losses post-disaster. However, the push for increased resiliency when rebuilding often gives way to the expediency of building it back faster and cheaper. This book is limited in exploring the intricacies of post-disaster reconstruction, and there are other books on recovery that cover this phase in greater depth and are referred to in the bibliography.

BASIC PRACTICES OF PROJECT PORTFOLIO MANAGEMENT

The purpose of project portfolio management is to optimize resources available through the management of different programs and projects to achieve the greatest benefits for the organization and the community it serves. This includes choosing a mix of programs and projects that enhance capabilities, evaluated against both the risks faced by the community, as well as the individual collection of program and project-based risks. The core practice areas of project portfolio management are Portfolio Life Cycle, Portfolio Governance,

Portfolio Capacity and Capability Management, Portfolio Stakeholder Engagement, Portfolio Value Management, Portfolio Risk Management, and Portfolio Strategic and Management.[11] These are discussed further below.

As portfolio management has similar practices found in both program and project management, let's walk through the practical methods of managing a portfolio that is different and how these relate to emergency management and public safety agencies.

- **Portfolio lifecycle management:** The core cycle of portfolio management is similar to that of a project, initiate-plan-execute, except the final stage is a portfolio review and optimization, i.e. balancing the risk-based return on investment in portfolio components. Agencies manage this process by applying a program and project-based methodology to managing these components as we have covered in the previous chapters and section of this chapter. It also looks at how it can improve its program/project methodology as it matures its practice.
- **Portfolio strategic management:** Identifying the strategic objectives of the organization. For example, an emergency management organization might look at:
  - the risk-based priorities (THIRA analysis) for the community or jurisdiction, and the gaps faced in managing them: prevention, mitigation, and preparedness response.
  - reviewing legal and regulatory requirements, e.g. adhering to the Disaster Recovery Reform Act of 2018, and engaging in projects that are funded by the Hazard Mitigation Grant Program and Pre-Disaster Mitigation program to mitigate wildfire or windstorms.[12]
  - political leaders' policies and priorities for public safety and security. An example of this would be an elected mayor whose campaign included building greater resilience to the effects of climate change.
  - consulting with community stakeholders, such as the private sector, non-profits, and community and faith-based groups, for their input into the strategic direction of the portfolio.
- **Portfolio governance:** Once consensus has been reached on strategic objectives, the next step is the processes for identifying, prioritizing, analyzing, and selecting a portfolio of component programs and projects that would be approved. This includes who and how decisions are made to approve the components. A sample methodology for project selection is provided in the following section.
- **Portfolio capacity and capability management:** Once portfolio components, programs, and projects have been selected, they are further refined based on organizational capacity and capabilities. Organizational capacities include staffing available to coordinate and collaborate on programs and projects, required resources (grant and/or tax levy funding, space, and equipment), and procurement bandwidth to handle the portfolio. Organizational capabilities to execute the portfolio include the competencies and attributes, e.g. management and technical knowledge, systems (IT and equipment), and the means to maintain and support the portfolio. For a long-term program to enhance public cyber security, an organization might look to grant funding that covers multiple years for development and a public-private partnership to sustain it in the out years. In addition, the operating organization needs to have the platform and staff to maintain and support any IT systems used for cybersecurity programming.

- **Portfolio stakeholder engagement:** Portfolio stakeholders are similar to project stakeholders in that they have a vested stake in the portfolio, with a degree of interest and influence. However, they differ, in that portfolio stakeholders are usually at a higher managerial level, executive, and decision-makers, who have an interest in the resource allocation and strategic outcomes associated with the portfolio, and less interest in the individual programs and projects[13]. As they are commonly senior elected leaders, agency chiefs, community leaders, etc., they have a strong proclivity for communication on portfolio progress, especially reports that demonstrate clear, measurable benefits to the public. Engaging portfolio stakeholders at the appropriate communication level and frequency is a key practice to gain and maintain their support for the portfolio.
- **Portfolio value management:** Portfolio value management is the realization and demonstration of measurable benefits to the community, similar to that in program management. As with program management benefits, portfolio value may be demonstrated in reducing community risks, enhancing hazard mitigation, and increasing emergency preparedness and response capabilities with clear indicators of positive improvements. Value management seeks to balance the portfolio's return on investment against the potential risks encountered.
- **Portfolio risk management:** Again, similar to program and project-based risk, portfolio risk management deals with managing the uncertainty in realizing the portfolio value in achieving the associated strategic objectives. This involves looking at the return on investment based on the portfolio stakeholders' relative level of risk appetite; risk appetite is the level of tolerance for uncertainty in achieving the expected value from the portfolio. The complexity of the portfolio helps to define the collective risks.

PORTFOLIO GOVERNANCE AND COMPONENT SELECTION

Too often we enjoy the comfort of opinion without the discomfort of thought. There are risks and costs to a program of action. But they are far less than the long-range risks and costs of comfortable inaction.

President John F. Kennedy, Speech at the opening of the
White House Conference on Aging on January 9, 1961

The key function of portfolio governance is deciding how to invest and manage the limited resources available to the organization. When selecting which components, programs and projects, might be incorporated into a portfolio, the most common methods are the classic "Return on Investment" and time to reach a break-even point for conducting a purely financial quantitative evaluation, which is appropriate when the financial return can be calculated. While some mitigation projects might lend themselves to these financial formulas, some project benefits cannot be easily quantified. In this case, there are two additional methods that can be combined for a group to use for evaluation:

- **Risk complexity analysis:** This method looks at the technical aspects of specific programs and projects and assesses different risk parameters for these components, beyond those in a standard probability and impact analysis of a risk rating matrix.[14] Since we are looking at the downsides of negative risks to components, opportunities or positive risks are not considered in this analysis. The risk complexity index seeks to apply an objective method of evaluation.

- **Multi-voting:** This method establishes a set of selection criteria for a portfolio that a group can vote on individually to create scores for each portfolio component that allows for prioritization. Criteria may include opportunities or positive risks and can be weighted. The multi-voting method is a subjective method of evaluation but is aimed at reaching a group consensus on component prioritization.

A risk complexity index starts with identifying several parameters for measurement and creating a scale for each, usually similar in measure, but they can also use weighted criteria according to its perceived value or impact. Then it is charted out in a grid and calculated according to the parameters against one another, similar to a risk rating matrix.
The following are steps creating a risk complexity index:

1. Identify several technical risk parameters to measure.
2. Create a scale for each risk parameter. If preferred, weights can be assigned to each scale to increase the rank of specific criteria.
3. Chart it out in a grid and score each portfolio component.
4. Calculate the parameters against one another.

The example in Table 14.1 analyzes three possible risk components, for three programs and two projects that might be commonly considered for an emergency management agency, looking at the complexity of three different criteria:

- **Technical complexity:** How technically difficult is it to execute the component and manage it? This includes staff expertise for maintenance and support once the component is deployed. The greater the difficulty and the less staff expertise, the higher the score.
- **Organizational risk:** How many staff members and time (direct from the organization, not grant-related) are involved, both in development and integration? The more staff and hours, the higher the risk score.

Table 14.1 Risk Complexity Index for New Portfolio Components

Program/Project	Technical Complexity (1–5)×8	Organizational Risk (1–5)×6	Contractor Network (1–5)×6	Total Score (max. 100)
Community Hazard Mitigation Education Program	16	18	6	40
Improve facilities and services to meet the needs of people with access and functional needs (DAFN Program)	32	30	24	86
Professional Development Program	24	24	18	66
UAV (Drone) Search and Rescue Project	40	30	30	100
Update ICS system	32	24	24	80

- **Contractor network:** How developed is the contractor network and how much experience is there with the component's work? The less developed the contractor network, the higher the score.

This simple example above shows that the UAV Search and Rescue Project has the greatest degree of technical risk, while the Community Hazard Mitigation Education Program has the least. This risk analysis can then be factored into the next step in evaluating these components.

MULTI-VOTING FOR SELECTING PORTFOLIO COMPONENTS

Multi-voting for portfolio components can provide transparency to the process of group decision-making. It can be helpful when the organization has a range of options that have to be narrowed down and when the decision must be made by group judgment.

The following instructions cover multi-voting for selecting portfolio components:

1. The first step is identifying evaluation criteria. As we covered in this section, this should be based on the organization's strategic objectives. For illustration purposes, we will use the following set of strategic objectives (simplified for this example) for the components given above:
   a. Improve community resilience by engaging the whole community in mitigation and preparedness by marshaling resources and taking actions to address critical vulnerabilities.
   b. Enhance the capability to provide situational awareness through the coordination and collection of essential information and providing critical information requirements for agencies and partners for effective decision-making during emergency incidents.
   c. Increase the capacity of the emergency management agency to quickly deploy and coordinate field resources to meet the life-safety mission.
   d. Improve the capability of the agency and its partners to provide services in an equitable manner to better support the communities served.
   Display the list of objectives and combine any duplicate items. Use an affinity diagram to eliminate any duplication and overlap.
2. Number all the items or use a code to identify them quickly and easily; for this example, we will use the alphabetical order as our code for these criteria.
3. Create weights for each criterion. This can be done by having each team member allocate 100 points among the criteria and averaging them. An example is provided in Table 14.2.
4. Score the projects based on the evaluation criteria using a similar scoring method. I recommend using a total of 100 points allocated by each team member among the criteria. An example of one project is given in Table 14.3.
5. Develop one chart with all of the weighted scores for all portfolio components, including a risk complexity index for reference. You may use a spreadsheet or document for this, any medium where you can put them a chart. A sample chart is given in Table 14.4.
6. Discuss the results to see whether there is a team consensus or if they are not clear.

Table 14.2 Multi-voting Criteria Weighting

| Criteria | Governance Team Member | | | | | |
	Hillary Folsom	Lucy Wang	Vikram Patel	Angelo Carnevali	Suzanna Maphalala	AVG.
A	25	30	20	20	20	**23.0**
B	15	30	10	20	50	**25.0**
C	35	20	50	30	20	**31.0**
D	25	20	20	30	10	**21.0**
Total	100	100	100	100	100	100.0

Table 14.3 Multi-voting Example for Program 1

| PROG1: Community Hazard Mitigation Education Program | Team Member Name | | | | | |
Criteria	Hillary Folsom	Lucy Wang	Vikram Patel	Angelo Carnevali	Suzanna Maphalala	AVG
A. Improve community resilience by engaging the whole community in mitigation and preparedness	80	60	70	50	60	**64.0**
B. Enhance the capability to provide situational awareness	0	0	0	20	10	**6.0**
C. Increase capacity to quickly deploy and coordinate field resources	0	0	0	10	10	**4.0**
D. Improve the capability to provide services in an equitable manner	20	40	30	20	20	**26.0**
Total	100	100	100	100	100	100.0

Table 14.4 Multi-voting Example for Portfolio Component Option

| Project Code | Portfolio Evaluation Criteria | | | | Total | Risk Complexity Index |
	A	B	C	D		
PROJ1	0.00	19.00	6.82	0.42	**26.24**	100
PROJ2	0.00	16.50	8.06	1.68	**26.24**	80
PROG3	2.30	8.00	10.54	4.20	**25.04**	66
PROG1	14.72	1.50	1.24	5.46	**22.92**	40
PROG2	2.76	0.00	4.96	15.12	**22.84**	86

The purpose of the discussion is to look at any dramatic voting differences and avoid errors from incorrect information or understandings about them. The discussion should not result and pressure on anyone to change their score one way or another. What you want to get at is why they scored a project in a certain manner and the

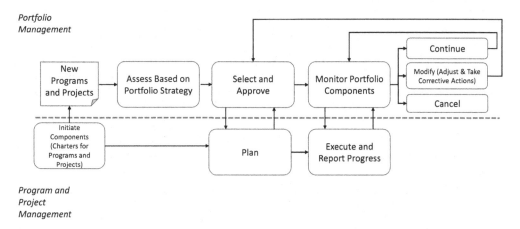

Figure 14.7 The portfolio management process. (Adapted from The Portfolio and Project Management Linkages;Kendrick, T. (2024). *Identifying and Managing Project Risk: Essential Tools for Failure-Proofing your Project.* American Management Association. p. 505.)

reasoning behind it. You may want to weigh the selection criteria against the risk complexity index. For example, PROJ1 UAV (Drone) Search and Rescue Project was given a weighted score of 26.24 but has a Risk Complexity Index of 100. In comparison, PROG1 Community Hazard Mitigation Education Program was given a weighted score of 22.92 with a Risk Complexity Index of 40. If resources are tight, then PROG1 and PROG3 might be given approval, while PROJ1 and PROJ2 might be scheduled for later when staff resources can be freed up and greater management support can be devoted to these projects.

As Figure 14.7 illustrates, portfolio management is a dynamic, continuous process that plans and controls the resource allocation among its components. The process involves reviewing where the organization needs to invest and disinvest based on changes in its strategic objectives, organizational capabilities, and funding sources.

Portfolio management begins with reviewing new programs and projects based on the initial charters and assesses them to see where they align with the strategy and selecting components. Program and project managers may adapt their plans to ensure that the end benefits suit the strategic portfolio objectives. Portfolio governance monitors the performance of the components as progress is reported and has managers continue as planned ("steady as they go") or take corrective actions and adapt components as necessary. In some cases, the governance team may allocate additional resources to support projects, or they may cancel the component if it not performing as planned or strategic objectives have changed. Ultimately, the portfolio results in benefits that are incorporated into the organization to update and enhance operational capacity.

Los Angeles Fire Department Strategic Plan for 2018–2020

The Los Angeles Fire Department (LAFD) is one of the largest municipal fire departments in the United States, employing close to 4,000 dedicated firefighters, paramedics, and support staff. LAFD works tirelessly to protect the 4 million residents and visitors of Los Angeles. With an operating budget of around $783 million in 2022 and equipped with advanced technology and training, LAFD firefighters respond to various emergencies, including fires, hazardous materials incidents, and natural disasters. LAFD paramedics provide advanced life support and emergency medical care. The department also emphasizes fire prevention, public education, and community outreach to promote safety and well-being.

The *LAFD a Safer City 2.0, Strategic Plan 2018–2020* laid out the department's strategic three-year plan. The plan lays out four major goals, strategies, programs, and projects to achieve the goals, as referred to in Figure 14.8.

Among the components of this strategic plan[15] were the following:

- A $15.5 million grant from FEMA to restore fire engines at four LAFD stations and to hire additional firefighters.
- A pilot program that provided Advanced Life Cycle Teams in the LAX (airport) terminals to support millions of travelers.
- A new SOBER (Sobriety Emergency Response) Unit, in partnership with the LA County Dept. of Public Health, to provide support to people experiencing homelessness who are struggling with alcohol abuse.
- LAFD High School Magnet Program that offers high school students training on firefighting and EMS skills and a career path to the LAFD.
- Development and implementation of a UAV program.

These strategic portfolio components have yielded benefits to LAFD and the Angelinos and visitors they serve. Although it is difficult to link improvements directly to the capabilities of the LAFD with end results, the City of Los Angeles experienced the lowest number of deaths due to structural fires in 2019.[16] While the overall number of service calls has steadily risen, average operational response times for EMS, Critical ALS, and structure fire, have all remained steady from 2019 to 2022.[17] In a community survey, 90% of respondents strongly of somewhat approved of the job LAFD is doing and felt that LAFD treats residents fair, courteous, and in a professional manner; over 80% of respondents answered that LAFD was very good or somewhat good at suppressing fires and responding to EMS calls.[18] LAFD a Safer City 2.0, Strategic Plan

GOAL

1

As we continually strive to provide exceptional public safety and emergency services, the Los Angeles Fire Department will enhance our customer service delivery model by leveraging performance management principles, innovative technology and adaptable resources to save and enrich lives, and to prepare for disasters – while fostering and enabling community resilience.

– ALFRED L. POIRIER,
CHIEF DEPUTY,
EMERGENCY OPERATIONS

LAFD provided structure protection at the Creek Fire where 115,000 residents were forced to evacuate their homes in December 2017.

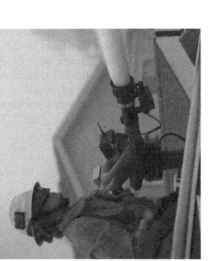

BRUSH FIRES

The 2017 California brush fire season produced an unprecedented amount of wildfires. The LAFD battled three historic brush fires locally. The La Tuna Canyon Fire erupted on September 1, 2017, and was the largest brush fire in 50 years at 7,194 acres burned. Just three months later on December 5, with erratic Santa Ana winds gusting at over 40 mph, the Creek Fire in Kagel Canyon broke out destroying 60 homes, 63 outbuildings and burned over 15,000 acres.

While many crews were battling the Creek Fire, the Skirball Fire ignited along the 405 Freeway, damaging 12 homes and destroying six. LAFD members engaged in an aggressive coordinated attack with partnering agencies to successfully limit the damages of many structures. These two simultaneous incidents depleted LAFD resources throughout the City; however, the Department-wide response was critical in preventing the loss of life, minimizing the destruction of structures and firefighter injuries.

LAFD strike teams were also deployed to other major wildfires including the Mendocino Lake Complex Fire in Northern California (36,523 acres, 545 structures destroyed), Thomas Fire in Ventura County (271,750+ acres, over 1,000 structures destroyed), Lilac Fire in San Diego and the Anaheim Hills Fire in Orange County.

STRATEGY 1: Improve emergency response times

- Continue implementation of Automatic Vehicle Location (AVL) technology and related Computer Aided Dispatch enhancements
- Enhance maps and other geographic information system related tools to improve situational awareness and public information
- Integrate with other systems, such as Automated Traffic Surveillance and Control and coordination of traffic lights
- Increase resource availability by providing transport to alternative patient destinations such as SOBER Centers and mental health clinics
- Expand Fast Response Vehicle (FRV) Program to provide increased coverage to areas of high call load and long travel distance

STRATEGY 2: Improve the delivery of Emergency Medical Services

- Develop and implement an EMS Strategic Plan
- Assess feasibility of implementing a firefighter/paramedic workload relief program
- Implement Paramedic Assessment resources at the remaining 24 Fire Stations, enabling the Department to be fully rotational
- Develop EMS Virtual Paramedicine Program that utilizes technology and social media platforms for minor patient assessments and treatment
- Expand Department of Mental Health Information Sharing Trial Program citywide to connect at-risk homeless individuals with housing and services
- Continue to expand Advanced Provider Response Units to have one deployed in each Battalion
- Assess feasibility of implementing Ambulance Operator Program

STRATEGY 3: Improve fire suppression services

- Develop and implement an Unmanned Aerial System Program
- Conduct Annual Field Incident Management Team Training
- Institutionalize after action reporting process for significant incidents
- Expand cadre of wildland qualified Command Officers
- Institutionalize a process for ordering and utilizing fixed wing aircraft
- Enforce "No Trespassing" Ordinance in Very High Fire Hazard Severity Zones to increase public safety and reduce risk of fires

Figure 14.8 LAFD a Safer City 2.0, Strategic Plan 2018–2020, p. 10.

NOTES

1 Project Management Institute. (2017). *A Guide to the Project Management Body of Knowledge* (PMBOK guide). p. 715.

2 Project Management Institute. (2017). *A Guide to the Project Management Body of Knowledge* (PMBOK guide). p. 714.

3 Project Management Institute. (2017). The Standard for Program Management, Fourth Edition.

4 Regional Partners www.regionalcatplanning.org/about_regionalpartners.shtml

5 RCPGP New York City & Northern New Jersey Area, www.regionalcatplanning.org/about_rcpgp.shtml

6 New York, New Jersey, Connecticut, and Pennsylvania Regional Catastrophic Planning Team (NY-NJ-CT-PA RCPT) Site Accomplishments, July 15, 2015. Note: this only reflects some of the programs and major projects, but not all of them.

7 New York, New Jersey, Connecticut, and Pennsylvania Regional Catastrophic Planning Team (NY-NJ-CT-PA RCPT) Site Accomplishments, July 15, 2015.

8 "2020 Population and Housing State Data". United States Census Bureau. Retrieved August 19, 2021.

9 Bay Area UASI FY 2020 – 2021 Annual Report Site, see: www.bayareauasi.org/news

10 Dinsmore, P. C., & Cabanis-Brewin, J. (2006). *The AMA Handbook of Project Management*. American Management Association. p.416.

11 Project Portfolio Performance Domains. Source: Project Management Institute. (2017). The Standard for Portfolio Management, Fourth Edition, pp.10–11.

12 FEMA Federal Insurance and Mitigation Job Aid for Disaster Recovery Reform Act, Section 1205 Additional Activities for Wildfire and Wind Implementation under Hazard Mitigation Assistance Programs December 3, 2019. See: www.fema.gov/sites/default/files/2020-07/fema_DRRA-1205-implementation-job-aid.pdf

13 Project Management Institute. (2017). The Standard for Portfolio Management, Fourth Edition, pp.63.

14 Kendrick, T. (2015). Identifying and Managing Project Risk: Essential tools for Failure-Proofing your Project. American Management Association.

15 Los Angeles Fire Department a Safer City 2.0, Strategic Plan 2018–2020, p.3–4.

16 Press Release, LAFD Spokesperson: Nicholas Prange LAFD SEES RECORD LOW STRUCTURE FIRE FATALITY DEATHS IN 2019, SETS NEW RECORD FOR NUMBER OF INCIDENT RESPONSE, Monday, January 13, 2020.

17 LAFD City Wide Response Metrics www.lafd.org/

18 Los Angeles Fire Department a Safer City 2.0, Strategic Plan 2018–2020, p.38.

Chapter 15

Emergency Response as a Project

(With) long-term projects, you have the ability to really plan more thoroughly and create more steps and …more of a thoughtful, staggered process, and then in an emergency you are really taking the steps, but putting it in a really compressed timeline. Coming up with the defined deliverables, data needs, analytics, and incremental steps help you say this is who we need to bring together to come to a consensus. We identify the issue, we come up with an analysis or review process. I mean all of that is built into the outcome that you desire. It's all project management.

Interview with Heather Roiter, Executive Director of Hazard Mitigation and Recovery,
New York City Emergency Management Department; 2019

Up to this point, this book has covered the application of project management that is focused on developing organizational capabilities to prevent, mitigate, and prepare for emergency incidents and disasters. This is what is called "Blue Sky" planning, done in advance when there is no major emergency incident on the immediate horizon and there is enough time to engage diverse stakeholders, conduct a thorough analysis, and develop the capabilities necessary. Now we examine the applicability of this methodology to plan during an emergency response and recovery, what is called "gray sky" or "black sky" planning. "Gray Sky" situations are when a disaster situation may be imminent or impacting,[1] and planning for response efforts is underway, and a "Black Sky" situation refers to an incident where power is completely out for a significant period of time and may impact other critical lifelines (food supply chain, housing, water, telecommunications, etc.).[2] These are incidents like major hurricanes and wildfires, high magnitude earthquakes, large-scale cyber terrorism, and a high-altitude electromagnetic pulse. Planning for gray sky and black sky incidents is what emergency planning and preparedness is all about, and when the incident triggers have been met, plans are deployed with all of the "staff, stuff, systems, and spaces" that have been prearranged.

As discussed in earlier chapters, the prime management methodology in the United States for responding to major emergency incidents is the Incident Command System (ICS) developed as part of the National Incident Management System (NIMS). ICS has the planning function embedded as a core of the methodology, and the Planning P is a representative model of how this planning works, with the "Operational O" as the planning cycle for every operational period (refer to Figure 15.1).

The incident action planning process synchronizes planning and emergency operational response, and it is often employed both in the field, at a command post, and at

DOI: 10.1201/9781003201557-18

Planning P with Project Planning Phases and Practice Areas

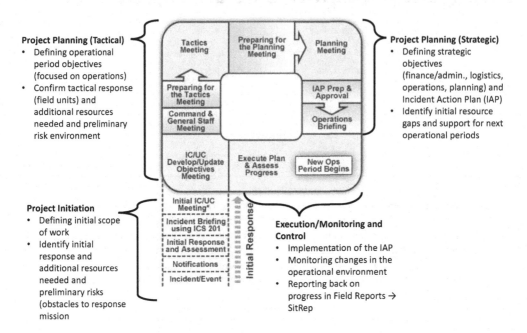

Project Planning (Tactical)
- Defining operational period objectives (focused on operations)
- Confirm tactical response (field units) and additional resources needed and preliminary risk environment

Project Planning (Strategic)
- Defining strategic objectives (finance/admin., logistics, operations, planning) and Incident Action Plan (IAP)
- Identify initial resource gaps and support for next operational periods

Project Initiation
- Defining initial scope of work
- Identify initial response and additional resources needed and preliminary risks (obstacles to response mission

Execution/Monitoring and Control
- Implementation of the IAP
- Monitoring changes in the operational environment
- Reporting back on progress in Field Reports → SitRep

Figure 15.1 Planning P. (From Incident Action Planning Process, ICS Course: FEMA IS 300: Intermediate Incident Command System for Expanding Incidents.)

an emergency operations center. The IAP establishes the "battle rhythm" and provides a framework to guide planning and decision-making. The process develops mutually agreed on operational objectives and how the emergency response structure will help support it. The IAP process is designed to facilitate coordination between different response agencies (at multiple levels) and partners, optimize the use of response resources, and avoid gaps and redundancies. It serves as a communications hub to "get the right information to the right people at the right time". This allows leadership, both elected leaders and senior emergency managers, to have a common operating picture and communicate expectations and provide clear policy guidance to incident managers. Incident managers in turn can receive the support they may need as it is available.

The IAP specifically:

- informs incident staff and partners of the incident objectives for the operational period and the operational environment, including the situation status, weather, and obstacles for response;
- indicates what resources have been marshaled for the response, including the specific type and kind, locations, and status of those resources;
- identifies actions that will be taken during the operational period to achieve the objectives;
- identifies work assignments and provides a clear action during the operational period;
- explains the organizational structure of the response, at the field and/or EOC level, and when and what type of planning meeting and briefings will take place.

The development of IAPs is a cyclical process, with personnel going through the planning steps repeatedly during each operational period until the incident is brought to a close and resources are demobilized. The cycle is represented by the Operational Period Planning Cycle (Planning O) part of the graphic above. When developing the IAP, planning staff utilize the most accurate and up-to-date information available at the time of the planning meetings.

Every emergency management agency has plans for emergency response, whether this is an Emergency Operations Plan (EOP) or Comprehensive Emergency Management Plans (CEMPs, designed for all hazards response) for a general response, or Emergency Action Plans (EAP) for response specific plans (sometimes included as annexes in the EOP or CEMP), which may be hazard specific, location specific, or function specific such as mass care, pandemic, and power outages. In addition to these plans, many jurisdictions which deal with frequent threats and hazards have standing objectives and resources ready for deployment, in particular in cases of a catastrophic event, e.g. massive earthquake, extreme hurricane, or tornado. These standing objectives may include the following (with measurable performance capacities for the jurisdiction[3]):

- establishing communications for critical responders (local, state, Federal/national);
- evacuation out of hazardous areas (hot zones);
- providing emergency shelter, water, and food to affected populations;
- initiate search and rescue operations when it is safe within a certain period of time; and
- provide medical assistance to injured people, and recovery and storage of decedents.

These plans describe the concept of response operations, identifying the resources, timeline, and costs associated with them and how they will be carried out, and these plans should be operational, e.g. have the resources available and means to execute (more on this later).

IS PROJECT MANAGEMENT NEEDED FOR RESPONSE?

If plans are in place which describe the scope, schedule, and budget for emergency response work, then, in essence, the key question to ask is whether project management methods are applicable during an emergency response, and first question we should answer is whether an emergency response fits the definition of a project. The answer to this is "yes"; a major emergency response is a project as it is "an undertaking involving the commitment of a significant investment of resources to produce a unique outcome in a specified amount of time", although with the following caveats:

- The timeline for response is not clear at the start until the magnitude of the emergency and its impacts have been assessed and how long response efforts will take.
- The scope of work is not completely clear as the critical needs are not immediately defined.
- The level of response resources (costs) won't be clear until the response is concluded; there might be some estimates at the start, but the full "bill" will not be definitive until all costs have been accounted for.

As we have laid out in the previous chapters, project management supports working cross-functionally, providing a framework for cooperation, coordination, collaboration, and communication while maintaining accountability to execute an end result, deliverable or outcome. While emergency management agencies and departments may structure it differently, within ICS, the planning function works in conjunction with operations, logistics, and finance and administration sections, provides support for short- and long-term planning, collects and issues situation and status reports, manages the resource assignments, documentation, and demobilization process. A sample ICS chart is given in Figure 15.2.

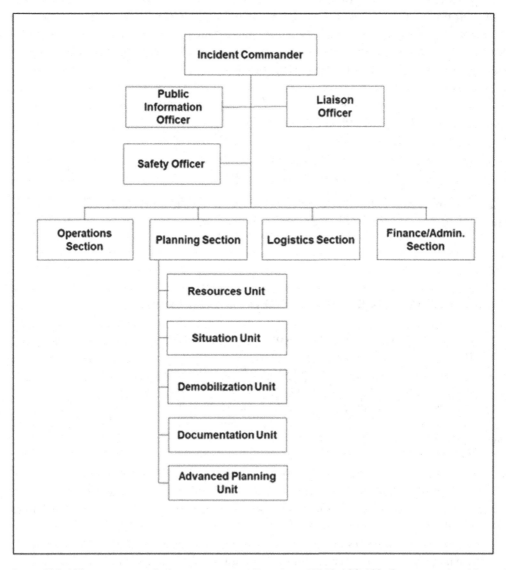

Figure 15.2 ICS structure and planning section. (Based on FEMA ICS 100 Course; with additional Advanced Planning Unit.)

By and large, this works for short to medium term responses, and in situations with limited complexity that are within the scope of the prescribed planned responses and available. Emergency plans and the response structures, for example EOCs, have activation levels that correspond to the type of event and describe what type of resource deployments are required. However, the planning function can break down in some of the following situations:

- when the scope of response is extended over a longer term than is expected;
- in a highly complex incident that exceeds the capacity of the response agencies and organizations (most commonly called catastrophic);
- in novel events, where the responsibilities of each agency in an incident are not clearly understood.

Some catastrophic plans provide for mutual aid and for the development of task forces that are designed to bridge the gap in planning, but quite often there is a lack of methodology and tools to support the type of coordination, cooperation, collaboration, and communication required for full leveraging this planning capacity.[4]

In principle ICS and NIMS were developed to address the dynamism of different types and sizes of emergency events by scaling to the level of response and providing a level of flexibility. When a disaster event exceeds the emergency response capacities of the entity (organization, governmental jurisdiction, etc.), the operating characteristics of the disaster environment are more dynamic due to following factors:

- Changes in the hazard/threat (impacts, duration, volatility, velocity, geography).
- Changes in the infrastructure (critical): Utilities – energy/power, transportation (air, rail, roads, water), food supply chain, telecommunications/internet, mass media, government services, social services, water/sewage, healthcare (DHS recognized 16 critical infrastructure areas).
- Changes to the responding emergency management organization; grows, shrinks, reorganizes supplemented or augmented by additional resources, elements are replaced and replenished (ICS org./Field); performance changes over time, degradation or enhancements.
- Changes to physical resources available: equipment, materials, supplies; vital supplies (food, water, medicine); information systems.
- Changes to the impacted population and how they respond to the impacts of the hazard and threats.
- Changes to the community's needs and behaviors.

MANAGING COMMUNITY EXPECTATIONS

One of project management's key responsibilities is managing stakeholder expectations throughout the project lifecycle. Similarly, during an emergency response, the community's expectations need to be managed, and community expectations will vary in a disaster situation, especially when the incident exceeds the scope of capacities or an expected timeline. This also depends on a number of factors, including the degree of community resilience

Figure 15.3 Maslow's Hierarchy of Needs.

and self-reliance, how well that community was functioning prior to the disaster, and extent of social capital.[5]

Beyond the standard models of Emergency Support Functions (ESFs)[6] or Community Lifeline Components discussed in Chapter 3, a broader framework to consider as part of meeting community needs and related expectations is based on Abraham Maslow's Hierarchy of Needs,[7] the Community Hierarchy of Needs,[8] referred to in Figure 15.3.

The following are descriptions and some examples of how the "Community Hierarchy of Needs" can be applied to a community impacted by a disaster incident:

- **Physiological needs**: A community meets the physiological needs of its members by providing access to food, water, shelter, and clothing. Providing for mass care (ESF 6) is one of the key emergency response functions.
- **Safety needs:** A community meets the safety needs of its members by providing a safe and secure environment. Depending on the type of incident, law enforcement, fire service, and EMS provide for security and search and rescue where needed.
- **Social needs:** A community meets the social needs of its members by providing opportunities for connection and belonging through neighborhood groups and associations, schools, faith-based and community-based organizations (CBOs/FBOs), sports teams, fitness clubs, local arts, cultural, and hospitality/entertainment spots and informal social connections. In a disaster, most of these organizations (non-profits included) and for-profits are responsible for their own response, continuity of operations, and recovery with limited support from governmental sources. In addition, community members attend to their basic needs before attending to social needs, although the community members often rely on those social networks for assistance during and in the aftermath of recovery.
- **Esteem needs:** A community meets the esteem needs of its members by providing opportunities for recognition and achievement. In the midst of a disaster, basic needs such as physiological and safety come first, so esteem needs are often pushed aside, although we may often see the media or formal groups acknowledge some individuals, first responders, officials, or unofficial "heroes" for the work they have done in the response.

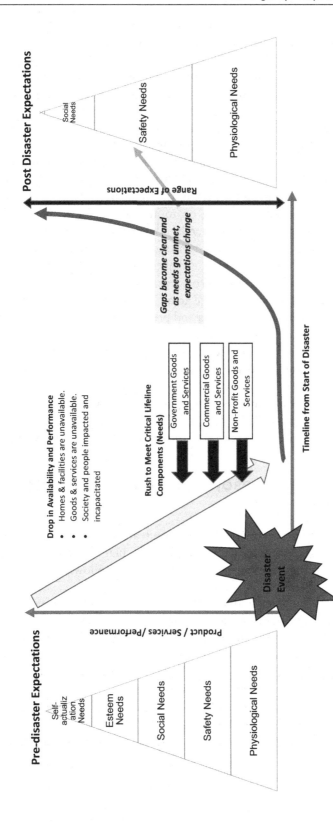

Figure 15.4 Channel of community expectations.

- **Self-actualization needs:** A community meets self-actualization needs of its members by providing opportunities for personal, as well as community, growth and development. As with esteem needs, self-actualization takes a back seat to basic needs in a disaster, although the response efforts may provide an opportunity for growth and development, often depending on the level of social cohesion prior to the event.

According to Maslow's model, the needs at the bottom of the pyramid need to be met before the needs above them can be met, and the process of meeting needs at the top is not always reached and, once met, is a continuous process of improvement and maintenance.

Refer to Figure 15.4. Because of the urgency to "stop the bleeding" in the aftermath of a disaster, the initial focus is on addressing the immediate basic needs, such as food, water, shelter, and security before moving on to any social needs. Indeed, in the aftermath of a catastrophic event like a tornado, rebuilding only begins once the community is "back on its feet" and can take care of itself to engage in the work of recovery, including reestablishing all previous levels of commerce, daily norms, and social connections through faith-based, community, and governmental organizations.

When emergency response capabilities are greatly unmatched to the needs, the longer it takes to address critical lifeline components and provide for basic community needs. This then results in gaps begin to become clearer to the community and contribute to growing expectations. However, community expectations are dynamic and diverse, depending on the community makeup and the gaps between pre-disaster and post-disaster expectations. It is also impacted to a degree by how those expectations are managed during the response, by elected officials and community leaders. This includes how efficiently the response efforts can fill those gaps by coordinating or adding resources or improvising new responses as required, as well as communicating with the community about what they can expect.

PLANNING DIFFERENCES: PROJECT TEAMS VS. EMERGENCY RESPONSE TEAMS

There are also clear differences between how project teams and emergency response teams operate that can influence longer term planning, in particular for disaster situations where scope and time may exceed expectations. These include the following:

- **Sense of urgency:** While preparedness and mitigation project planning allows for time to engage stakeholders and elaborate the plan before executing, emergency planning and response requires immediate solutions and actions. This involves intense, hyper-focused efforts to analyze and identify solutions to meet critical needs but can also create challenges in identifying, allocating, and coordinating necessary resources effectively.
- **Stakeholder groups:** While project plans for preparedness have defined stable stakeholder groups, emergency responses often involve a wide range of stakeholders that can be dynamic, including government agencies, local organizations, and affected individuals and communities.

- **Team dynamic:** Preparedness project teams have the opportunity and time to become familiar with one another and to become an integrated team, but they often have limited authority over prescribed resources. Emergency planning teams responding to a disaster are often professionals derived from different disciplines and organizations but are empowered with the authority to develop viable solutions, request, and marshal resources as approved.

- **Risks are different:** Emergency response involves higher stakes as the mission prioritizes life-safety, safety for first responders, and the public they serve, and then protecting physical buildings and other assets from damage. This requires a certified level of training and equipment, e.g. PPE. Preparedness and mitigation project planning is generally lower stakes for the risks faced; projects can be course corrected as they are developed or cancelled, usually with minor risk to the project team, stakeholders, or the end-users served.

However, in spite of these, there are opportunities in assigning an advanced planning team that adapts and incorporates project management methods into the planning process and can help both expedite and elaborate effective solutions. A small, multifunctional advanced planning team made up of qualified and experienced individuals from a subset of ICS organization, i.e. Operations, Logistics, Finance/Admin., and professionals with functional expertise would be included. Now that we have covered the approximate scope and potential team makeup, we turn our attention to the potential planning methodology.

ADVANCED AND DYNAMIC PROJECT MANAGEMENT METHODS

As we have pointed out above, the ICS planning process does not currently accommodate detailed planning for large sets of related tasks that do not fall within the operational planning period or outside the scope of a standard emergency response. While this work is the domain of EM planners who address gaps in long-term plans or in response scope, there is no clear doctrine that governs these planning needs. In principle, the EOP, CEMP, or EAP would address these needs, but these planning processes are a static preparedness planning methodology which are ill-suited to the vicissitudes of disaster environments.

However, deterministic project management methods may be inadequate for these planning needs as they assume fixed inputs with a set process and fixed outputs. There are other project planning methodologies, and the following chart illustrates a proposed mix of existing project planning methods that may address these planning needs (refer to Table 15.1).

The project management method used for planning needs to suit the circumstance and organizational culture. In some cases, certain methods can also be flexibly applied to discrete elements required for planning and not the whole project, e.g. a deliverable or decision point. These methods are not necessarily applied exclusively and may be used in tandem in certain situations. While this book has been devoted to the application of standard project management methods (deterministic methods based on the PMBOK©), we are not going to cover these other methods in great depth. However, we will now briefly explain these methods to see how they can be applied to the world of emergency management and public safety and under what circumstances they might be utilized.

Table 15.1 Emergency Management Planning Needs Mapped to Potential PM Methods

Project Planning Method	EM Planning Description	Purpose
Rolling Wave Approach	Planning for the immediate period that can be defined and estimating resource needs and scheduling for that given period, and leaving detailed planning until the next period approaches and planning considerations are clear.	Visibility in forecasting the future needs and requirements allows for planning for the immediate horizon and then as the project is executed shift planning that as we move along. This can be used as part of standard project management.
Agile Project Management (also referred to as Scrum/ Kanban)	Iterative and flexible approach to managing projects that emphasizes plan adaptability, close collaboration with the customer to quickly complete functional units of work output.	Originally developed for software application development in contrast to standard project management covered in this book, it addresses changes in requirements and task prioritization in complex and rapidly changing environments.
Program Evaluation and Review Technique (PERT)	A probabilistic planning method that uses three point estimates, optimistic, and pessimistic, and most likely, for the purpose of providing a range of different time estimates for each task and for the project as a whole. Similar to standard project management, but provides a greater level of analysis.	Useful when potential ranges exist in inputs/ resources (people, equipment, materials/ supplies, systems), anticipated demand for resources and services in the population served (survivors/beneficiaries).
Critical Chain PM (CCPM)[12]	CCPM is a version of PERT which uses ranges of estimates with the difference between the pessimist time range and the probable estimate become a buffer time that is added to the end of a task, or milestone size, or the end of the project. The task owner manages the task duration, and the PM manages the buffer time.	Changes in time estimates due to variations in resources, demands on services and supplies, and operating conditions.
Graphical Evaluation and Review Technique (GERT)	GERT is similar to PERT, although it incorporates elements of decision trees with conditional time and resource estimating adding decision loops for some task considerations.	Changes in operating conditions (weather, safety, security, inputs, process, population, threat/hazard, capacities) and in strategic or tactical options that might be pursued.

ROLLING WAVE APPROACH

The Rolling Wave Approach is a project management technique that involves progressively planning and executing a project in waves, with detailed planning and execution carried out for the near-term activities based on immediate requirements that can be clearly defined, while leaving the future activities at a higher level of abstraction until the scope of work is clearer. This approach acknowledges that project details and requirements may evolve over time and that attempting to plan everything upfront may not be practical or efficient. Two diagrams are presented below that explain this planning concept in Figures 15.5 and 15.6.

Here are the key steps of the Rolling Wave Approach:

1. **Iterative planning:** Instead of planning the entire project in detail from the beginning, the Rolling Wave technique focuses on planning and executing activities in shorter time frames or waves. These waves are typically a few weeks or months long, depending on the project's duration.

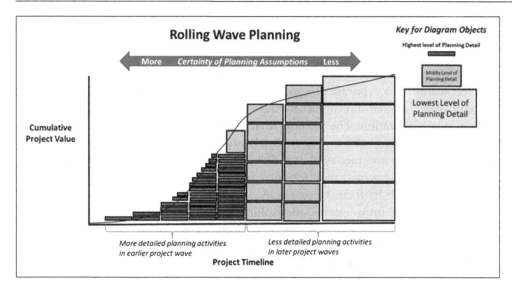

Figure 15.5 Rolling wave planning for project management – First Wave.

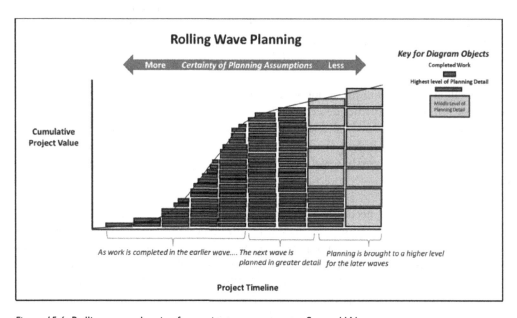

Figure 15.6 Rolling wave planning for project management – Second Wave.

2. **Detailed planning for current wave:** The activities and tasks in the current wave are planned in detail, including defining objectives, deliverables, resources, and timelines. This detailed planning allows for more accurate estimation and allocation of resources. This is similar in fashion to what has been presented in the book thus far.
3. **High-level planning for future waves:** While the current wave is planned in detail, future waves are planned at a higher level of abstraction. These waves are outlined

with broader objectives and milestones, allowing for flexibility and adaptation to changing project conditions.

4. **Continuous refinement:** As each wave progresses, the Rolling Wave technique allows for continuous refinement and adjustment of the project plan. This flexibility enables project teams to incorporate new information, address emerging risks, and align the project with changing stakeholder needs.

5. **Progressive elaboration:** The Rolling Wave technique embraces the concept of progressive elaboration, where project details are developed and refined over time as more information becomes available. This iterative process helps in managing uncertainty and evolving project requirements.

6. **Risk management:** By focusing on detailed planning for the current wave and high-level planning for future waves, the Rolling Wave technique allows for early identification and mitigation of risks associated with the known activities. Risks and uncertainties for future waves can be captured as part of the high-level planning and addressed in subsequent planning iterations and project execution.

The Rolling Wave Approach offers several benefits, such as increased adaptability to changing project conditions, improved accuracy in planning and estimation, and efficient resource allocation. However, it requires close collaboration and communication among project stakeholders to ensure alignment and effective management of project waves. Note that the Rolling Wave Approach is often used in conjunction with other project management methodologies, such as Agile or Scrum, to provide a structured framework for iterative planning and execution.

The Rolling Wave technique, along with elements of Graphical Evaluation and Review Technique (GERT) (see further description below), was largely what I used during my first project assignment, the Medical Assistance Project in Croatia between 1993 and 1995. Due to the lack of health data available on the target population served by the project, refugees, and displaced persons, we undertook a parallel track of project activities with various levels of detail:

- Conducting a rapid needs assessment to define what specific pharmaceuticals were needed. This involved visiting over 100 health clinics and hospitals serving refugees and displaced persons, explaining the project and what the requirements were. It also included meeting with the World Health Organization (WHO) to coordinate the project work with similar NGOs providing for healthcare needs of the same population.
- Establishing a transportation network and tracking and reporting system for the pharmaceuticals delivered to medical professionals serving the target population.
- Training our staff on our project goal and distribution, monitoring, and reporting procedures.
- Establishing partnership agreements with hospitals and health clinics to cover administration and reporting for pharmaceuticals to the target population.
- Initiating procurement for pharmaceuticals.

As the first procurement was completed and pharmaceutical products were deployed to our network of partner clinics and hospitals, we started with the next level of detailed planning, reassessing needs, coordinating with the WHO and other NGOs, and improving our distribution and tracking capabilities.

AGILE PROJECT MANAGEMENT

Agile Project Management is an iterative and flexible approach to managing projects that emphasizes adaptability, collaboration, and delivering early on product value for the customer. It was originally developed as a response to the limitations of traditional project management methods in handling complex and rapidly changing environments, in particular software application development, where customers may have latent requirements which may not be completely clear at the start of a project. The diagram below illustrates the Agile Project Management process using the Scrum model; the term Scrum was derived from the rugby scrum where progress in bringing the ball up the field may not always be linear.

Refer to Figure 15.7. In Agile Project Management, projects are broken down into smaller, manageable units called development "sprints" or "iterations". Each development iteration typically lasts for a fixed duration, often ranging from one to four weeks. At the beginning of each iteration, the project team and stakeholders collaborate to identify the most valuable features or tasks from the burndown list to work on during that period.

The key elements of Agile Project Management methodology include the following:

1. **Iterative development:** Work is divided into short iterations, allowing for continuous feedback, learning, and adaptation. The iterations are work sprints that deliver units of work to the customer or end-user for testing and feedback.

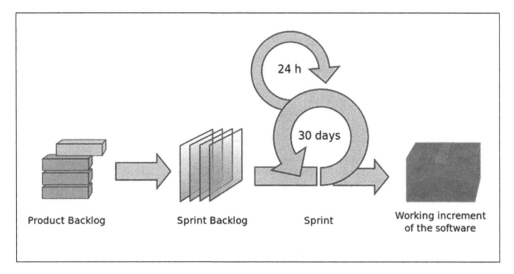

Figure 15.7 Agile Project Management process: Scrum model. (Source: "Scrum process" by Lakeworks – Own work. Licensed under GFDL via Commons – https://commons.wikimedia.org/wiki/File:Scrum_process.svg#/media/File:Scrum_process.svg.)

2. **Cross-functional teams:** Multidisciplinary teams work collaboratively, including members with diverse skills and expertise, promoting effective communication and shared ownership of the project's success.

3. **Customer collaboration:** Regular engagement with customers and stakeholders helps ensure that the project's direction aligns with their needs and expectations. At the end of each sprint, the work is shared with the client for feedback and corrections and changes, and a reprioritization of the burndown list (requirements).

4. **Adaptive planning:** Rather than creating a detailed plan upfront, agile projects focus on creating a high-level roadmap, elaborating a set of requirements into a burndown list, and adjusting it based on changing requirements, new information, and feedback received during each iteration based on customer feedback.

5. **Continuous improvement:** Emphasizing learning and adaptation, Agile Project Management encourages the team to reflect on their progress, identify areas for improvement, and make necessary adjustments to enhance team productivity and efficiency.

Agile Project Management methodologies, such as Scrum and Kanban, have strict frameworks and guidelines for implementing agile principles effectively. These methodologies offer specific practices, ceremonies, and tools that enable teams to organize their work, track progress, and deliver value incrementally throughout the project lifecycle. One tool that is commonly used on agile projects is the Kanban (translates from the Japanese as

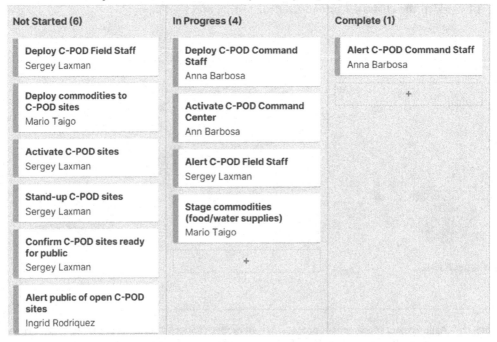

Commodity Points of Distribution (C-POD) Deployment Plan Kanban

Not Started (6)	In Progress (4)	Complete (1)
Deploy C-POD Field Staff Sergey Laxman	**Deploy C-POD Command Staff** Anna Barbosa	**Alert C-POD Command Staff** Anna Barbosa
Deploy commodities to C-POD sites Mario Taigo	**Activate C-POD Command Center** Ann Barbosa	+
Activate C-POD sites Sergey Laxman	**Alert C-POD Field Staff** Sergey Laxman	
Stand-up C-POD sites Sergey Laxman	**Stage commodities (food/water supplies)** Mario Taigo	
Confirm C-POD sites ready for public Sergey Laxman	+	
Alert public of open C-POD sites Ingrid Rodriquez		

Figure 15.8 Project planning Kanban example.

billboard or bulletin board, in other words planning board), shown below in an example based on the deployment of commodity points of distribution (C-POD) sites (refer to the example in Figure 15.8).

Agile Project Management works well in managing tasks with loosely defined completion times, but which have an urgency in completion. EOCs and command posts may benefit from using this methodology, and many ICS software applications use this framework for managing "tickets" or requests that come in, but when managing a series of tasks for a function-specific plan deployment, a separate system may be more flexible and helpful for reporting. There are a number of software applications that fulfill the functionality of a Kanban such as Jira, Trello, Miro, Asana, the MS Teams Planning Board, and Smartsheet.

PROGRAM EVALUATION AND REVIEW TECHNIQUE (PERT)

The Program Evaluation and Review Technique (PERT) is a project management methodology that helps in planning, scheduling, and coordinating tasks within a project. It was first developed in the late 1950s by the US Navy's Special Projects Office and the management consulting firm Booz Allen Hamilton.[9] PERT was initially used for managing large-scale defense projects, such as the development of missile systems. PERT has been used in industries and project types with complex and uncertain timelines.

PERT utilizes network diagrams to represent the sequence of activities and their dependencies within a project. The network is commonly represented as a graph, with nodes representing activities and arrows representing the flow and dependencies between them.

PERT largely follows the same techniques of critical path method (CPM) schedule development, including activities and precedence relationships between tasks. However, PERT differs from CPM with the following planning aspects:

- **Time estimates:** PERT typically incorporates three values for time estimates: optimistic, pessimistic, and most likely. These estimates capture the range of possible durations for an activity, taking into account best-case, worst-case, and expected scenarios for each task (see formula below).
- **Probabilistic analysis:** PERT employs probabilistic analysis to estimate the project's duration, taking into account the range of possible activity durations and their dependencies. This analysis provides a more realistic understanding of the project timeline, accounting for uncertainties and variations in activity durations.

The PERT formula used for probabilistic estimating is as follows:

Task Duration = (Optimistic Estimate + 4 × Most Likely + Pessimistic Estimate)/6

PERT may be best applicable when estimating a total time for a series of similar activities performed in parallel with range variations for each unit. For example, using the example above for C-POD deployments, if we estimate that standing up 20 C-POD sites will each take between 8 and 16 h with most likely at 10 h and the deployment of commodities from the warehouse will take between 24 and 37 h, with 32 h most likely, to complete, but the deliveries finish when the last C-POD has been stood up and has commodities available, our calculations for these two tasks would be as follows:

Task	Optimistic	Most Likely	Pessimistic	Duration
Stand up 20 C-POD sites	8	10	16	10.7
Deploy commodities to 20 C-POD sites	24	32	37	31.5

The planning estimates given above can be used in two ways. If we assume that deliveries can be made to C-POD sites, even with a small number of staff on site, then we can begin deliveries 8 h after C-POD sites have begun standing up. If our EOC leadership has asked for a time to be able to confirm C-POD sites are confirmed ready to open and receive the public, with confirmation taking perhaps 2 h, then we provide an estimate that sites will be available at 42 h from the start of this task series (8 h + 32 + 2) or in less than two days. Of course, there may be the expected push from elected leaders to get the job done more quickly. In that case, the delivery of food commodities might be accelerated by asking facility staff to open the loading areas and have someone available, which can then reduce the opening time to roughly 34 h (or 32 if leadership can live without confirming that they are ready).

CRITICAL CHAIN PM

Critical Chain Project Management (CCPM) is a project management methodology developed by Dr. Eliyahu Goldratt in the 1990s. It aims to address the challenges of resource constraints, task dependencies, and project schedule reliability that arise from a traditional project management approach. Like PERT, the schedule development includes activities and precedence relationships between tasks, but it takes a different approach to scheduling.

Instead of using a weighted average for task duration estimates, it uses the optimistic estimates and then provides for buffer time, which are managed between the team members and project manager. CCPM operates with an understanding of Parkinson's Law[10] that work takes up the volume of time that it is given and seeks to remove this artificial inflation of duration estimates.

Here are the fundamental elements of CCPM:

1. **Resource constraints:** CCPM recognizes that limited resources can hinder project performance. It emphasizes the need to identify and manage resource constraints effectively. This involves identifying the critical resources required for project completion and ensuring their optimal utilization.
2. **Task estimates:** CCPM recognizes that task owners, those responsible for the duration estimate, want to ensure they provide enough time to finish a given task, so will often overestimate the time in order to protect themselves from being punished for delays. Instead, CCPM encourages task owners to be aggressive and uses optimistic estimates. Buffer time is created for each task (see below).

3. **Critical Chain:** It takes into account both task dependencies and resource constraints. By focusing on the Critical Chain (what would be called the critical path through the network), CCPM aims to reduce project duration and improve delivery reliability.

4. **Feeding chain and task dependencies:** CCPM emphasizes the identification and management of dependencies between tasks and their impact on the project schedule. Feeding chains refer to the series of tasks that feed into the Critical Chain, and their timely completion is crucial for maintaining the project schedule.

5. **Buffer management:** CCPM utilizes buffer management to address uncertainties and variations that can impact project schedules. Two types of buffers are used in CCPM: project buffers and feeding buffers. Project buffers are strategically placed to protect the entire Critical Chain from disruptions, while feeding buffers are used to protect dependent chains within the network. Task owners are accountable for meeting the task duration, while the project manager tracks and controls the buffers and releases time as needed to task owners when requested.

6. **Resource leveling:** CCPM promotes resource leveling, which involves balancing resource usage across the project tasks to avoid overloading or underutilizing resources at given times. By effectively managing resource availability, CCPM aims to reduce resource conflicts and optimize overall project performance.

An example of how CCPM can be visualized is given in Figure 15.9.

It can be used to manage multiple projects using the same scheduling principles. In a multi-project environment, it can be effective in addressing resource allocation and prioritization to ensure efficient and effective project execution. It's important to note that CCPM may require a mindset shift and adoption of new practices compared to traditional project management approaches. Successful implementation involves a reorientation to this methodology, a sense of trust and collaboration among project stakeholders, and

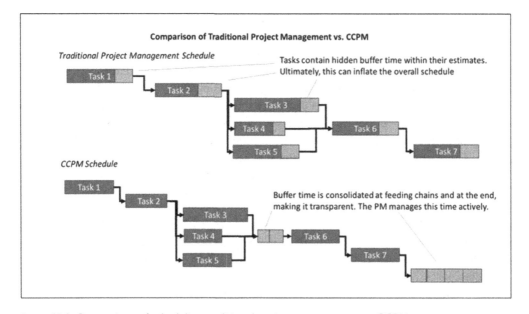

Figure 15.9 Comparison of scheduling traditional project management vs. CCPM.

support from leadership, as well as minor adaptations to project management software to suit the methodology.

GRAPHICAL EVALUATION AND REVIEW TECHNIQUE (GERT)

The GERT is a project management tool that extends the capabilities of traditional project network methods, such as the CPM or PERT. GERT was primarily developed in the 1960s at the Lockheed Missiles and Space Company, a major defense contractor at the time, under the sponsorship of the DoD's Advanced Research Projects Agency (DARPA).[11] GERT allows for modeling and analysis of complex projects that involve uncertainty, probabilistic events, and decision points.

GERT utilizes network diagrams to represent the flow of activities and events in a project. The nodes in the network represent events or decision points, while the arrows or lines represent activities or tasks, which is different from CPM. GERT introduces additional features to the network, such as probabilistic branches and conditional points, which make it suitable for handling uncertainty and iterative project flows.

Here are the key elements of GERT project scheduling:

1. **Events:** Events are represented by nodes as specific points in time within the project. They can have either deterministic (fixed outcomes and time durations, circle shaped) or probabilistic (multiple outcomes with ranges of time).
2. **Activities:** Activities are listed on arrows and represent the work required to be performed to move from one event to another. They have associated time estimates, durations, and dependencies.
3. **Probabilistic branches:** GERT allows for modeling probabilistic events that can occur during an activity. These events may impact the duration or outcome of an activity and are represented as branches in the network.
4. **Conditional branching:** GERT allows for modeling decision points where the path of the project depends on certain conditions or outcomes. These decision points are represented as branches with conditions or probabilities.
5. **Time estimates:** GERT uses time estimates for activities and events, which can be deterministic (fixed) or probabilistic (range of possible values). Probabilistic estimates help account for uncertainty and provide a more realistic view of project timelines.
6. **Simulation and analysis:** GERT enables simulation and analysis of project timelines to determine the critical path, project duration, probabilities of meeting deadlines, and other performance measures. This analysis helps identify areas of risk and assists in making informed decisions for project scheduling and resource allocation.

GERT uses activity on arrow with events expressed as nodes describing completed stages of development. A simplified example is given in Figure 15.10, using GERT for the evaluation and selection of a new mass notification system. This system includes both deterministic and conditional activities. Completed events for this schedule are given below:

1. Start (dummy node, used for illustrating the project start)
2. Evaluation of the current mass notification system

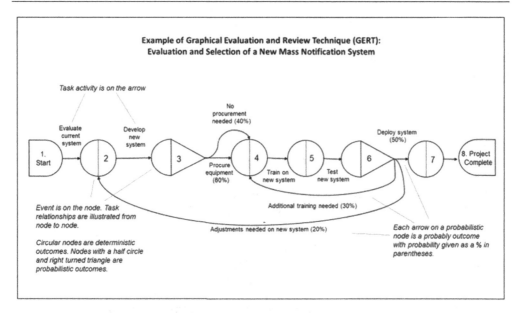

Figure 15.10 Example of graphical evaluation and review technique (GERT).

3. Development of new mass notification system
4. Procurement of equipment for the new system
5. Staff training for the new system
6. Testing the new system
7. System deployment
8. Project complete (dummy node, used for illustrating the project end)

For instance, depending on the new system requirements (3), the procurement of equipment (4) may be needed or may not be needed: 40% probability is not needed, and 60% procurement is needed. Also, the amount of training completed (5) depends on the success of testing the new system (6), need for additional training (30%) and need for additional adjustments to the new system (20%), with a 50% probability that testing is successful, and the system can be deployed and finish deployment (7). Lastly, task durations can be estimated using both deterministic and probabilistic methods, similar to PERT, and a Monte Carlo simulation run to see the range of potential outcomes.

Despite its intricate methodology, GERT provides project managers with a more comprehensive and flexible method for analyzing project schedules involving uncertainty, probabilities, complex dependencies, and decision-making. It allows for a deeper understanding of project dynamics, potential risks, and helps in creating more robust project schedules.

WHAT PLANNING METHODS MIGHT BE APPLIED NOW?

Some of the methods presented above may be applied in place of or in addition to the standard project management methods covered previously. Some methods are easier to implement. Agile planning methods have a large number of software solutions. However, adopting the agile planning methodology requires training, experience, and client flexibility in applying it for the first time. It may also require contingency resources to remedy any schedules that go off the rails.

The Rolling Wave approach is more of a planning strategy as it can be integrated with standard project management methods and does not require additional software. The same is largely true with Critical Chain as it only requires adding a buffer time into the schedules. PERT estimating can be accommodated by some software programs, but some require manually calculating these estimates, e.g. MS Project, Smartsheet; working with it can become quite complex as the number of tasks increases. When used on a larger scale, GERT requires software applications that can accommodate conditional estimating, so few project planning software applications support it; although there are some advanced, specialized tools in systems engineering, many of them designed for complex industries. Fundamentally, in applying any planning method, the question that needs to be asked is whether the benefits outweigh the cost of using it.

PLANNING FOR DEMOBILIZATION AND TRANSITION TO RECOVERY

If every project comes to end, then when and how does that end occur when an emergency response is a project? From a project perspective this is what we call closure, as we discussed in Chapter 13, Monitoring and Control: Tracking Project Progress. Under ICS this is called demobilization, and it happens when all resources have been effectively demobilized and the situation in the ground has stabilized. In larger scale emergency responses, the recovery process often follows demobilization. While the demobilization process is often planned and carried out, where demobilization ends and recovery begins is a relatively gray area.

The chart in Figure 15.11 provides a simplified version of the demobilization process (presented as the "Planning Q"). The process starts once the emergency situation begins to stabilize and response operations begin to wind down. Although some emergency planners start developing a demobilization plan, an exit strategy, after the initial IAP, planning usually starts at the point when a trigger has been reached and a decision to demobilize has been made once the situation on the ground has begun to stabilize. The plan needs to take into consideration of what and where resources are deployed, timing for demobilization, and how recovery will be handled and to whom this will be transitioned. Responsibility for recovery in some jurisdictions and instances may or may not fall to the emergency management agency involved in the initial response. It is beneficial to have an organization responsible for recovery (whether this is a part of the responding organization or external to it) and a recovery plan (or semblance of one) to transition recovery responsibilities. The situation continues to be closely monitored for any changes, and if it has begun to destabilize, then a decision may be made to return to the ICS planning process cycle.

The demobilization plan will also include some of the following elements:

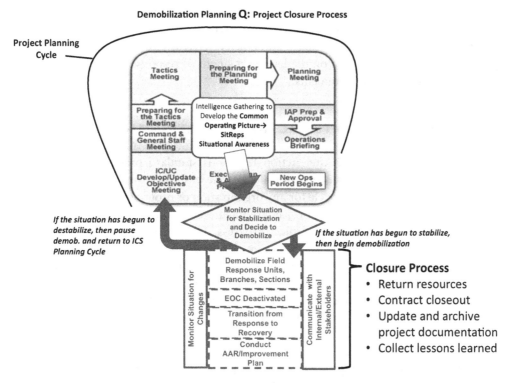

Figure 15.11 Demobilization Planning Q. (From Incident Action Planning Process, ICS Course: FEMA IS 300: Intermediate Incident Command System for Expanding Incidents; Planning Q addition to Planning P developed by the author in conjunction with All Hands Consulting. Adapted and used by permission.)

- **Communications:** A communications plan, informing response staff and coordinating agencies and partners when demobilization will begin and conclude (in some cases, a range with an approximate date to finish), and demobilization progress as it is implemented.
- **Resource disposition:** Resource reassignment or return to its original status. Equipment is returned to storage or warehouse, reconditioned and repaired when necessary. Supplies are restocked if they have a long shelf life or inventoried for rotation for future emergency deployments.
- **Documentation:** All final reports are written up and all documentation collected for archiving.
- **Debriefing:** One of the final steps before demobilization is engagement in a hotwash and a more comprehensive After-Action Review, to evaluate the emergency response as a whole, and development of an Improvement Plan (IP; see below).
- **Recuperation and restoration:** Staff need time to recover from an emergency response (rest and relaxation) before they return to other duties. This should also include providing for psychosocial needs to process the critical incident experience.

In effect, the demobilization process should be looked at as a project plan in itself, with a start and finish, albeit with some degree of uncertainty on exactly when these points are met. The ultimate goal is to orderly and efficiently return all resources to a state of readiness (as possible depending on the level and length of response) and to evaluate the response to make future improvements.

DEBRIEFING AND THE AFTER-ACTION REPORTING AND IP

The practice of conducting an After-Action Review is standard practice following any major emergency response. This is the equivalent of the lessons learned process and is part of the continuous improvement cycle for any organization or program. We covered project closure and how the outcomes of a project are integrated into operations in Chapter 13, Monitoring and Control: Tracking Project Progress. The formal process of conducting a final briefing, collecting lessons learned from the deployment, culminates in a written After-Action Report and Improvement Plan (AAR/IP). While it has the characteristics of a project, we will not go into great detail on the AAR/IP process, as it is defined in Homeland Security Exercise and Evaluation Program (HSEEP) under Module 6 Evaluation. The fundamental questions answered in an AAR/IP are similar to those in a lesson learned:

- What happened?
- What was supposed to happen (according to plans, SOPs, etc.)?
- What gaps were there between the two and how can improvement be made: plans, resources, training, etc.?

Notice that I avoided the words "what went wrong". This is intentional as the AAR/IP needs to engage all levels of people involved in the emergency response and should be conducted in as transparent manner as possible. This requires an organizational culture that is fault tolerant (within responsibilities and according to SOPs or leadership intent). The value of important lessons laid bare in an open discussion after the heat of intense critical responses is often cleaned up and covered over, lest they embarrass higher ups in the organization. The AAR/IP process can be easily politicized, and there can be a tendency to avoid the scrutiny that might come with making it as open and transparent, so it is often helpful to have a document that "for official use only" for internal consumption vs. one released to elected officials. There is also a strong dichotomy between what can be changed in response and what may fall outside the ability to improve. We will explore the continuous improvement cycle and quality in emergency and public safety management in the next chapter.

NOTES

1   Red Alert: The Official Blog of the American Red Cross North Texas Region, *five Lines of Service: Disaster Relief*, By Shannon Randol, intern contributor, American Red Cross.
2   Schnurr, A. (2018). Black Sky Hazards: Systems Engineering as a Unique Tool to Prevent National Catastrophe. In: Madni, A., Boehm, B., Ghanem, R., Erwin, D., Wheaton, M. (eds) *Disciplinary*

Convergence in Systems Engineering Research. Springer, Cham. https://doi.org/10.1007/978-3-319-62217-0_69

3 For example, a city the size of San Francisco, CA might plan for food and water to up to 500,000 survivors following a magnitude 9 earthquake.

4 ICS was developed based on some concepts of military "command and control" (C2) organizational structures, but in today's emergency management environment, there is often a diverse group of governmental, private, and non-profit organizations, and at different levels, involved in the response, so these "top-down" principles are ill-suited. The use of the term C3 standing for coordination, cooperation, and collaboration appears to be more appropriate.

5 Aldrich, D. P. (2019). *Black Wave: How Networks and Governance Shaped Japan's 3/11 Disasters*. University of Chicago Press. pp.14–16

6 Emergency Support Functions (ESFs) provide the structure for coordinating Federal interagency support for a Federal response to an emergency incident. The ESF model has been adopted by some state and local emergency management agencies to distribute the work in an EOC. See www.fema.gov/emergency-managers/national-preparedness/frameworks/response#esf

7 Maslow, A. H. (1943). A Theory of Human Motivation. *Psychological Review*, 50(4), 370–396. https://doi.org/10.1037/h0054346

8 *Localist: a Concept3D Company, The Community Building Hierarchy of Needs* by Myke Nahorniak, November 20, 2020, www.localist.com/post/the-community-building-hierarchy-of-needs.

9 Malcolm, D. G., J. H. Roseboom, C. E. Clark, W. Fazar. (September–October 1959). Application of a Technique for Research and Development Program Evaluation. *Operations Research*, 7(5), 646–669.

10 Parkinson, Cyril Northcote (19 November 1955). Parkinson's Law. *The Economist*. London.

11 Kannan, R. (2014). Graphical Evaluation and Review Technique (GERT). *Advances in Secure Computing, Internet Services, and Applications*, 165–179. https://doi.org/10.4018/978-1-4666-4940-8.ch009

12 Based on the Critical Chain Theory by Eliyahu Goldratt. See also www.pmi.org/learning/library/critical-chain-project-management-investigation-6380

Chapter 16

Quality Management in Emergency Management Programs and Continuous Improvement for Responses, Projects, and Programs

> It is not enough to do good. Good must be done well.
> *Denis Diderot, 18th Century French philosopher, writer, and encyclopedist*

Most public safety and emergency management organizations, including international humanitarian agencies (also called relief organizations), have a formulized set of guidelines, standards, and better practices associated with the program and projects. They are designed to promote continuous improvement. In this chapter we will explore some of the quality methods and tools and how they may be applied to quality improvement of emergency management and public safety programs (not projects; this was covered in Chapter 16). We will identify some of the common standards and where they are applied, and the use of exercises and the conduct of an AAR/IP as part of the lessons learned and improvement process.

GUIDELINES FOR DEVELOPMENT OF EMERGENCY PLANS, SYSTEMS, TRAINING, AND EXERCISES

Figure 16.1 illustrates the relationship between the Plan-Do-Check-Act (PDCA) Quality cycle attributed to W. Edwards Deming,[1] combined with the emergency preparedness cycle propagated by FEMA. The planning phase is followed by the "doing": equipping (staff, stuff, spaces, and systems), training and exercises and response (real life; Note – I have added this to the diagram); then checking how well it performed, followed by acting again to make improvements with a return to planning. With this in mind, using a standard or guidelines provides a reference when developing emergency and public safety management capacities through projects and programs.

FEMA has a number of standards and planning guides for developing plans, training, and exercises, and standards for building emergency management capabilities discussed earlier in this book, including CPG 101, HSEEP, and the 32 Core Capabilities embodied in the National Preparedness Goal. The United Nations office for the Coordination of Humanitarian Affairs (OCHA) is responsible for coordinating emergency responses for the various UN Agencies and works through the Inter-Agency Standing Committee (IASC). IASC publishes guidelines for emergency preparedness and response. There are

DOI: 10.1201/9781003201557-19

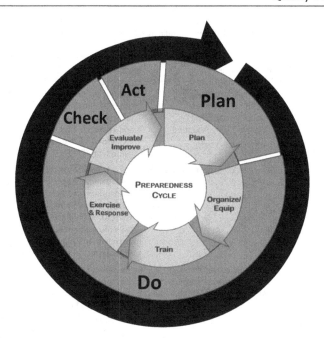

Figure 16.1 PDCA and FEMA preparedness cycle.

also a number of other planning guides for hazard mitigation (disaster risk reduction according to the UN) at the state and international level. I recommend consulting these guides as a first step when creating a development methodology and defining the scope and set of requirements for projects and programs.

Table 16.1 includes those standards and some additional basic guidelines. Note – this list of planning and development guidelines is not intended to be comprehensive or exhaustive. There are additional function or hazard-specific planning and development standards and guides available that are not covered here.

These standards are revised periodically as better practices are informed by field experience and updated. In the same manner organizational emergency and public safety management practices need to be reviewed, updated, maintained, and supported, lest they decay. This is why updates to plans, training, and exercises are so important in the absence of organizational experience in real-life emergency response. As Figure 16.2 illustrates, this is the core of a continuous improvement for any organization and their standard operating procedures and guidelines. An organization builds up their capabilities based on a core of better practices and steadily makes improvements over time.

QUALITY METHODS AND TOOLS

Some of the quality methods and tools we covered in Chapter 9, Developing the Quality Plan, are also suitable for evaluating the operational capabilities in emergency management and public safety. Along with the standards given above, some of these methods may be applied to public safety and emergency management programs as follows:

Table 16.1 Emergency Management and Public Safety Standards (Basis of Common US Emergency and Public Safety Management Practice)

Planning Standard/Guideline	WWW Reference	Application
California Adaptation Guide, June 2020, Cal OES, Governor's Office of Emergency Services	https://resources.ca.gov/ CNRALegacyFiles/ docs/climate/01APG_ Planning_for_Adaptive_ Communities.pdf	This guide is "designed to help local government, regional entities, and climate organizations to incorporate best practices and current science and research into their adaptation plans". This is a California State guide similar to the THIRA/SPR guide from DHS FEMA below.
CALEA® Standards for Law Enforcement Agencies	www.calea.org/node/11406	The Commission on Accreditation for Law Enforcement Agencies, Inc. (CALEA®), a credentialing authority through the joint efforts of law enforcement's major executive associations. CALEA Standards cover major law enforcement functions, ethics, training and exercises, organizational structure, management, and administration for law enforcement agencies.
Community Lifelines Implementation Toolkit: Comprehensive information and resources for implementing lifelines during incident response, v.20 November 2019 (US DHS FEMA)	www.fema.gov/emergency- managers/practitioners/ lifelines-toolkit	As described in the toolkit, lifelines provide an outcome-based, survivor-centric frame of reference that assists responders in assessing the scale of a disaster and to develop operational response priorities and objectives, and transitioning to recovery outcomes. "Lifelines describe the critical services within a community that must be stabilized or re-established – the ends – to alleviate threats to life and property". Emergency Support Functions (ESFs) and the Core Capabilities help to address the steps to "stabilize or reestablish lifelines".
Developing and Maintaining Emergency Plans Comprehensive Preparedness Guide (CPG) 101, September 2021, v.3.0 (US DHS FEMA)	www.fema.gov/sites/default/ files/2020-05/CPG_101_ V2_30NOV2010_FINAL_ 508.pdf	Framework for developing and maintaining emergency plans at the Federal, state, tribal, and local levels. Primarily targeted to governmental agencies, but may be applicable to for-profit and non-profit organizations.
Emergency Management Standard, Emergency Management Accreditation Program (ANSI/EMAP EMS 5-2019)	https://emap.org/index. php/what-is-emap/the- emergency-management- standard	Framework for continuous improvement of emergency management programs, including planning, resource management, training, exercises, and evaluation. This standard is used as evaluation tool for accrediting emergency management programs at the state, tribal, and local levels.
FEMA Operational Planning Manual, FEMA P-1017, June 2014 (US DHS FEMA)	https://emilms.fema. gov/is_2002/media/ 142.pdf?msclkid= 22772168a6d811ec811ea 85b21732346	Planning guideline for FEMA's operational planning for both preparedness and incident response at the Federal and regional level, maximizes interoperability within FEMA and with other Federal partners, and transitions plan to operational and field units.
Homeland Security Exercise and Evaluation (HSEP; US DHS)	www.fema.gov/emergency- managers/national- preparedness/exercises/ hseep	"Provides a set of fundamental principles for exercise programs, as a common approach for program management, design and development, conduct, evaluation and improvement planning".
Incident Action Planning Guide, Revision 1, July 2015 (US DHS FEMA)	www.fema.gov/sites/default/ files/2020-07/Incident_ Action_Planning_Guide_ Revision1_august2015.pdf	Standard for incident action planning during FEMA incidents, but also used as a framework by state, tribal, and local governments.
Local Mitigation Planning Handbook, March 2013, (US DHS FEMA)	www.fema.gov/sites/default/ files/2020-06/fema-local- mitigation-planning- handbook_03-2013.pdf	Provides guidance to local governments in "developing and updating a local hazard mitigation plan". This relates to the THIRA/SPA and California Adaptation Guide but is specific.

Table 16.1 (Cont.)

Planning Standard/Guideline	WWW Reference	Application
National Disaster Recovery Framework, Second Ed., June 2016 (US DHS FEMA)	www.fema.gov/sites/default/files/2020-06/national_disaster_recovery_framework_2nd.pdf	Framework for "coordinating recovery assistance" and "supports local governments by facilitating problem solving, improving access to resources, and fostering coordination among state, tribal, territorial and federal agencies, nongovernmental partners, and stakeholders" to focus on community recovery needs.
National Incident Management System (NIMS) Third Ed., October 2017 (US DHS FEMA)	www.fema.gov/sites/default/files/2020-07/fema_nims_doctrine-2017.pdf	NIMS is the guiding doctrine for (emergency) incident management in the USA and governs how Federal agencies operate and integrate at the state, tribal, and local levels. NIMS and its subsidiary doctrine of ICS have been widely adopted at all levels of government throughout the United States. Note – some state and local governments have their own versions that integrate with NIMS, e.g. Standardized Emergency Management System (SEMS used on California), Citywide Incident Management System (CIMS used in New York City).
The National Preparedness Goal and Core Capability Development Sheets (US DHS FEMA)	www.fema.gov/national-preparedness-goal	The National Preparedness Goal, including the Core Capabilities (35 across 5 mission areas), is intended to develop "a secure and resilient Nation across the whole community to prevent, protect against, mitigate, respond to, and recover from the threats and hazards that pose the greatest risk".
NFPA®1,600 Standard on Disaster/Emergency Management and Business Continuity Programs 2019 Ed.	www.nfpa.org/assets/files/aboutthecodes/1600/1600-13-pdf.pdf	"Widely used by public, not-for-profit, nongovernmental, and private entities on a local, regional, national, international and global basis, NFPA 1600 has been adopted by the U.S. Department of Homeland Security as a voluntary consensus standard for emergency preparedness, including continuity and crisis management". This standard is often used as a reference for private and non-profit organizations. Note – the National Fire Protection Association develops a variety of codes and standards that govern fire prevention and suppression, emergency medical services, safety, and security. See: www.nfpa.org/Codes-and-Standards/All-Codes-and-Standards/List-of-Codes-and-Standards
Public Health Emergency Preparedness and Response Capabilities: National Standards for State, Local, Tribal and Territorial Public Health, Oct. 2018, Centers for Disease Control and Prevention, Center for Preparedness and Response	www.cdc.gov/orr/readiness/capabilities/index.htm	Standard designed to advance emergency preparedness and response capacity of state and local public health systems.
Target Capabilities List: A companion to the National Preparedness Guidelines, Sept. 2007 (US DHS)	www.fema.gov/pdf/government/training/tcl.pdf	The Target Capabilities List (TCL) is a comprehensive inventory of capabilities that jurisdictions and organizations should aim to achieve within their emergency management programs. It provides specific details and performance objectives for each capability area. Note – The FEMA Core Capabilities provide a high-level overview of critical emergency management functions, while the Target Capabilities List offers a more detailed inventory and performance objectives.

(Continued)

Table 16.1 (Cont.)

Planning Standard/Guideline	WWW Reference	Application
This is UNDAC, United Nations Disaster Assessment and Coordination United Nations Disaster Assessment and Coordination (UNDAC) Field Handbook	www.unocha.org/sites/uno cha/files/UNDAC_broc hure_2018_0.pdf https://www.unocha.org/ sites/dms/Documents/ UNDAC%20 handbook%20- %20English.pdf	"The UNDAC system is designed to support national governments, the UN in-country, Humanitarian Coordinators and Humanitarian Country Teams, and incoming international responders with coordination during the first phase of a sudden-onset emergency. It also aims to advise and strengthen national and regional disaster response capacity". The UNDAC Field Handbook guides UNDAC Teams in the field.
Threat and Hazard Identification and Risk Assessment (THIRA) and Stakeholder (SPR) Preparedness Review Guide, Comprehensive Preparedness Guide (CPG) 201, 3rd Ed., May 2018 (US DHS	www.fema.gov/sites/ default/files/2020-04/ CPG201Final20180525. pdf	The THIRA/SPR is a methodology to assess community risks (threats and hazards), identify and evaluate capabilities needed to respond to risks, and identify gaps in existing capabilities and set targets to meet. This is a part of the National Preparedness System and is intended to involve the "Whole of the Community". This is process oriented and relates to the Core Capabilities which are specify individual capabilities.
A Whole Community Approach to Emergency Management: Principles. Themes, and Pathways for Action, FDOC 104-009-1 December 2011 (US DHS FEMA)	www.fema.gov/sites/default/ files/2020-07/whole_ community_dec2011_ _2.pdf	A framework for "engaging with members of the community as vital partners in enhancing the resiliency and security of our Nation through a Whole Community Approach. and how to integrate the Whole Community concepts into daily practices" of emergency preparedness, response, and recovery.

International and United Nations Standards

Planning Standard/Guideline	WWW Reference	Application
Cluster Coordination at Country Level IASC Sub-Working Group on the Cluster Approach and the Global Cluster Coordinators' Group July 2015, Inter-Agency Standing Committee, UN	https://interagencystanding committee.org/ system/files/2020-11/ Reference%20Module%20 for%20Cluster%20 Coordination%20at%20 Country%20Level%20 %28revised%20July%20 2015%29.pdf	This UN-based guideline "outlines the basic elements of cluster coordination and intends to serve as a reference guide for field practitioners to help facilitate their work and improve humanitarian outcomes". The Cluster System is intended to help coordinate humanitarian aid by UN, international non-governmental agencies, and governmental agencies in response to a disaster incident. Note – UN has separate protocols for humanitarian assistance in refugee situations.
IASC Reference Module for the Implementation of The Humanitarian Programme Cycle, v.2.0, Humanitarian Programme Cycle Steering Group, July 2015, Inter-Agency Standing Committee, UN	https://interagencystanding committee.org/system/ files/2020-11/IASC%20 Reference%20Module%20 for%20the%20 Implementation%20of%20 the%20Humanitarian%20 Programme%20Cycle%20 2015%20%28Version%20 2%29.pdf	Provides a planning framework for international humanitarian response in disasters that do not involve "Defines the roles and responsibilities of international humanitarian actors and the way that they interact with each other, national and local authorities, civil society and with people affected by crises".
ISO 22301:2019 Security and resilience – Business continuity management systems – Requirements ISO 22313:2020 Security and resilience – Business continuity management systems – Guidance on the Use of ISO 22301	www.iso.org/standard/ 75106.html https://www.iso.org/ standard/75107.html	An international standard for certification that "specifies requirements to implement, maintain and improve a management system to protect against, reduce the likelihood of the occurrence of, prepare for, respond to and recover from disruptions when they arise… applicable to all organizations, or parts thereof, regardless of type, size and nature of the organization".

Table 16.1 (Cont.)

Planning Standard/Guideline	WWW Reference	Application
ISO 22320:2018 Security and resilience – Emergency management – Guidelines for incident management, International Organization for Standardization	www.iso.org/obp/ui/ #iso:std:iso:22320:ed-2:v1:en	An international standard for certification that "provides guidance for organizations to improve their handling of all types of incidents, e.g. emergencies, crisis, disruptions and disasters… targeted to organizations and agencies, with the private sector, regional organizations, and governments".
Sendai Framework for Disaster Risk Reduction 2015-2030, 2015 United Nations	www.undrr.org/publication/ sendai-framework-disaster-risk-reduction-2015-2030	"The Sendai Framework for Disaster Risk Reduction 2015-2030 outlines seven clear targets and four priorities for action to prevent new and reduce existing disaster risks: (i) Understanding disaster risk; (ii) Strengthening disaster risk governance to manage disaster risk; (iii) Investing in disaster reduction for resilience and; (iv) Enhancing disaster preparedness for effective response, and to 'Build Back Better' in recovery, rehabilitation and reconstruction". This UN framework is targeted to an international audience and relates to the UN Sustainable Development Goals.
Emergency Response Preparedness (ERP) October 2014: Risk Analysis and Monitoring; Minimum Preparedness Actions; Advanced Preparedness Actions; and Contingency Planning, 2015, DRAFT FOR FIELD TESTING, Inter-Agency Standing Committee, UN	www.humanitarianresponse. info/sites/www. humanitarianresponse. info/files/documents/ files/Emergency_ preparedness_guidance-24Oct2014.pdf	Provides practical guidance to assist humanitarian country teams in preparing to respond to potential emergencies with appropriate humanitarian assistance and protection.

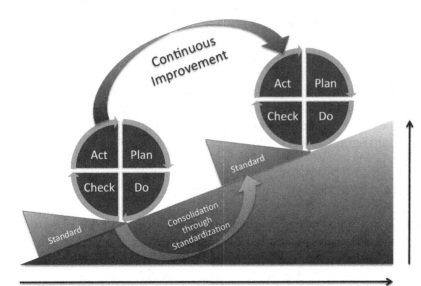

Figure 16.2 Continuous improvement cycle. (From Johannes Vietze https://commons.wikimedia.org/ wiki/File:PDCA_Process.png. This file is licensed under the Creative Commons Attribution-Share Alike 3.0 Unported license.)

Strengths	**Weaknesses**
-Good interpersonal relationships among the companies -Size of system makes it easy to work with and know other EMS/fire providers	-Age of providers (advanced) -No career opportunities -Training requirements to provide EMS are an obstacle for volunteerism -Unwillingness to adapt to changing medical standards -Attrition of certified EMS personnel – multiple causes -No formal administrative hierarchy for fire/EMS -Slow internet

SWOT Analysis

Opportununities	**Threats**
-Availability of grant funding sources -County budget has surplus. -Develop marketing/PR strategy plan -A credentialing authority could be established. - Improved utilization of manpower; North and South sharing manpower -To improve quality of members skills and interpersonal skills -Dedicated fill-time EMS coordinator -Background investigation before being allowed on apparatus	-Commissioners may not accept appropriate level of responsibility. -Lack of motivation and dedication -Paid service taking over volunteer service -Apathy – for the improvement process. Recommendations may not be implemented or may be watered down. -The plan meets real world and may not be viewed as a consensus document. -County government may have more control of volunteers. -Loss of county funding for support

Figure 16.3 SWOT analysis for EMS. (Adapted from the Garrett County (MD) Emergency Medical Services SWOT Task Force, Final Report February 23, 2006. www.garrettcounty.org/resour ces/emergency-services/pdf/swotreport.pdf.)

SWOT Analysis

SWOT analysis is helpful for strategic analysis of organizational capabilities. Some examples might be reviewing potential areas where programs and projects might be undertaken to address weaknesses or pursue opportunities. See an example in Figure 16.3 from a County Enterprise Management Systems (EMS) organization in Maryland:

DMAIC (SIX SIGMA APPROACH)

Six Sigma is a statistical process control methodology, first developed at the US Company Motorola to support manufacturing process improvements. The "Six Sigma Way", often called Six Sigma Design, is a standard process to methodically solve a problem effectively. It follows a similar process as the PDCA cycle and consists of the following steps from the pneumonic DMAIC:

- **Define the problem**: This step defines the problem to be solved. An example from the SWOT analysis above would be to address the problem of "Training requirements as an obstacle for voluntarism"; the goal of a potential project would be to redevelop training to overcome obstacles to recruiting and keeping volunteers.
- **Measure key aspects of the problem:** This step identifies key factors that may cause the problem and collects related and relevant data. This provides a quantifiable picture of the problem. Using the same problem above, we might conduct a survey of new volunteer recruits, including those dropping out, and find out what challenges they face persisting as an EMS volunteer. This might be supplemented by interviews. Gathering data may include the use of check sheets, described further below.
- **Analyze the data:** This step organizes and analyzes the data to identify trends and highlight areas to address. There are a few methods that we provide below such as the Ishikawa (Fishbone Diagram) and Pareto Analysis, discussed further below, although other quality analysis methods may also be effective.
- **Improve**: This step examines the current process (or if there is no process) for areas of improvement and the plan to improve based upon the data analysis conducted. This is where project planning comes into play and also sets the measurable improvement targets to be achieved.
- **Control**: this last step monitors and controls the new process and tracks any deviations from the measurable improvement targets. It identifies any potential factors that contribute to process deviations. Controls may include check sheets, statistical process control, inspections to continuously monitor the process. Improvements are implemented to the process until the targeted quality output is obtained. This may include starting over at the beginning of the process (continuous improvement).

While the statistical process control method that looks at standard deviations (hence the name Six Sigma[2]) may be helpful for very large data sets, in the tens of thousands or greater, for lower numbers, using a basic quality analysis process and other tools given below is sufficient.

CHECK SHEETS

Check sheets are often the basis of how data is collected, and this includes more sophisticated information systems, e.g. ICS applications, Enterprise Management Systems (EMS), and database-integrated software applications. A check sheet lists each different instance and number of instances, usually over a period of time, for example this may be errors in performance or requests to a 911 hotline during a disaster. A checklist example is provided in Table 16.2 for a 311 Call Center supporting residents during a flood event. Note – this is only for illustration and is not based on real data.

PARETO ANALYSIS

Pareto Analysis is a data analysis used to help identify the most important factors from a set of data in order to address the most important factors first. A Pareto Analysis would

Table 16.2 Check-Sheet Example for a 311 Call Center

DATE	Query on Evacuation	Query on Location for Shelter	Request Evacuation	Question on ADA Access	Pet Needs	Other	Total
7/14/21	459	393	0	12	56	23	943
7/15/21	736	512	126	123	89	12	1,598
7/16/14	418	235	278	183	123	37	1,039
7/17/14	0	24	34	96	32	6	168
Total	1,613	905	438	414	300	78	3,748

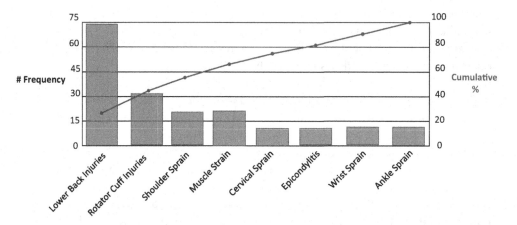

Figure 16.4 Annual workplace injuries-EMS operation. (Adapted from A Methodology for Detecting Musculoskeletal Disorders in Industrial Workplaces Using a Mapping Representation of Risk, November 2018. DOI:10.5772/intechopen.81710 License Creative Commons BY 3.0.)

be a common method to evaluate data collected from a check sheet. The Pareto Principle is also known as the 80/20 rule, as it is often shown that 80% of effects or defects come from 20% of the causal factors. Data is usually represented in a Pareto Chart, a type of bar chart where values are plotted in decreasing order of frequency from left to right, and a line is also drawn to represent the cumulative percentage of the total. This visual representation helps teams focus on the issues that will have the greatest impact when solved. The example provided in Figure 16.4 illustrates various injuries in an EMS operation over a year. In the case below, we might focus on providing assistive devices for EMS staff to aid in lifting as well as physical training on lifting techniques to help reduce the incidence of lower back and shoulder-associated injuries.

ISHIKAWA DIAGRAM (FISHBONE DIAGRAM FOR A ROOT CAUSE ANALYSIS)

We covered the use of the Ishikawa Diagram, often referred to as a Fishbone Diagram, in Chapter 9. It is a form of root cause analysis and developed as a graphical tool used to brainstorm potential causes that lead to the main problem explored. The technique of the 5 Whys (asking the question why five times) until all root causes have been identified, and

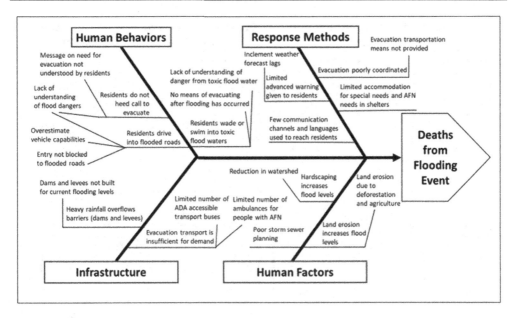

Figure 16.5 Fishbone diagram (Ishikawa diagram) for deaths from a flooding event.

displaying the potential causes of a specific effect. An example of an Ishikawa Diagram is provided in Figure 16.5 for the root causes of deaths from a flooding event.

DATA ANALYTICS AND PREDICTIVE MODELING

Data analytics and predictive modeling play an increasingly vital role in the practice of emergency management and public safety. Data analytics can help create predictive models to improve hazard mitigation, enhance emergency preparedness, streamline response efforts, and aid in recovery efforts. Whether it is used by internal departments or is outsourced to specialized third parties, it involves the use of advanced software and methodologies to collect, process, and interpret data from various sources such as UAVs (drones), earthquake and seismic data, weather reports and satellite imagery, social media, and emergency call records.

For example, the U.S. Geological Survey (USGS) uses data analytics and predictive modeling to develop earthquake hazard maps. These maps show the probability of an earthquake occurring in a given area and uses these maps to develop earthquake preparedness plans for communities in high-risk areas. Figure 16.6 provides an illustration of this analysis for a seismic event.

Data analytics and predictive modeling are powerful tools that can be used to improve our understanding of earthquakes and to reduce the risk of damage and loss of life from seismic impacts. As our understanding of earthquakes continues to grow, these data analytics and modeling tools will continue to evolve and become even more effective at enhancing community resilience to earthquakes.

During an ongoing emergency, real-time data analytics can track the incident's progression, enabling authorities to allocate resources effectively and make informed decisions.

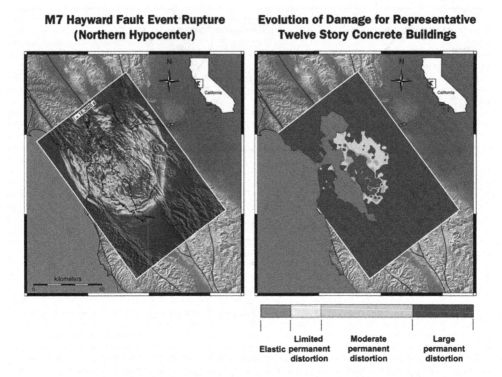

M7 Hayward Fault Event Rupture
(Northern Hypocenter)

Evolution of Damage for Representative
Twelve Story Concrete Buildings

Limited   Moderate   Large
Elastic permanent   permanent   permanent
distortion   distortion   distortion

Figure 16.6 EQSIM, advanced new earthquake simulation software, showing the Hayward Fault in the San Francisco Bay Area. The simulation illustrates both seismic wave propagation through the ground (left) and corresponding evolution of building damage (right). (From Berkeley Lab, Next Generation Earthquake Simulations for the Bay Area. See: https://youtu.be/m6Sp 6qaWuLs.)

Post-disaster, data analytics aids in assessing the damage, identifying the most affected areas, and planning reconstruction efforts. By integrating machine learning and AI, data analytics in emergency management is becoming increasingly sophisticated, improving the effectiveness and efficiency of disaster response and planning.

GENERAL GUIDANCE ON DESIGN AND DEVELOPMENT PLAN JOB AIDS AND TRAINING

I have picked up a number of lessons learned over the course of 20 years involved in the development of numerous job aids (derived from emergency and public safety documents) and training, in various modalities: classroom based, online (synch. and asynch.), and in simulated virtual worlds. The following better practices are provided with two caveats: there are specific better practices that can be found at associations (Association for Talent Development [ATD]) or in books on the subject; this is not intended to be comprehensive, and there will be situations where the ability to control the development within these guidelines may be limited. At times sharing the risk or opportunity or escalating the issue under a risk plan may be necessary.

Here are the guidelines:

- When developing training make sure to do a job analysis first. If developing this across a class of jobs, e.g. supervisory/managerial, then conduct a job analysis for each major one and consolidate this into one to develop the key competencies: knowledge, skills, and abilities.
- Before proceeding with development, make sure to get the complete base documentation when it is available: procedures, job descriptions/job action sheets, manuals, etc. in hand. Confirm that the documentation is the latest and that no additional changes will be made. If the documentation is subject to change, then you want to raise this as a potential risk and discuss how to address it (see Chapter 10 on Developing the Risk Plan).
- In addition to documentation, make sure that you lock down other content that may be used, i.e. multimedia, job aids, graphics, and forms. You will want to confirm who is providing this and whether the performing organization will be using stock media (purchased and/or free). Keep in mind a budget for purchased or newly created media.
- Make sure quality specifications required for the media are communicated to the client and adhered to.
- Make sure all content is documented (licensed content, royalty, or license free) and make sure you get the base electronic files for any reuse or the need to relicense in other courses; this is not just the SCORM package but also a repository of all those digital assets in their native form. Make sure that all subcontracted content, e.g. audio narration, is a work-for-hire arrangement (through the ICA).
- When converting a classroom-based course to an online course, consider having an early walk-through of the course using PPT slides and an instructor before any development begins. Consider recording the video or audio of the session. The investment of a couple thousand dollars may be worth the time spent on any rework.
- When developing training videos, use sample videos from stock vids to mimic what might be used on screen. Be very clear on the tone of delivery and word usage.
- Rigorously track all changes and ensure traceability from the design document through development stages, e.g. Alpha, Beta, Gamma development.
- If you are developing training courses, keep the number of courses you develop at any one time limited to two to three, depending on the length and complexity of the content. Ensure that you allow yourself and the client enough time to review courses as they are produced (see Chapter 14: Program and Project Portfolio Management); as above, make sure to track those changes carefully and how many rounds of edits (enforce these rules ruthlessly).
- Consider how you will test the designs or early prototypes, e.g. via walk-throughs, pilots, and end-user tests. If using a pilot, you need to answer the following questions:
  - How will you pilot the course(s) and how many pilots do you plan to conduct?
  - Who/How will you run the debrief?
  - How will the developer (and client) be involved?
  - Will you have a technical pilot before (and separate) from the content pilot?
- For online courses, make sure to test along the way, for example, ensure that the course is recording properly in the learning management system (LMS), and how

it performs SCORM calls, e.g. every 30 seconds, minute, 5 min, or at the end of a module, whatever is required. Confirm that the LMS tracks this properly and records completion. Are there assessments built into the course and do they allow tracking of performance on these and each individual test item and answer stem? If using an LMS, make sure to allow adequate time to test any new means of user registration electronically.

INSTITUTIONAL AND COMMUNITY KNOWLEDGE OF DISASTERS

There are many challenges in maintaining and supporting community and institutional knowledge of disasters and emergency response. Once a disaster event has subsided and recovery has ended with the community, there is often a rush to return to a sense of normal routine that there is often a tendency toward wanting to forget what happened. The steady drift toward complacency ignores the measures and resources built up to address the original disaster. A good example of this is the Great Influenza Pandemic of 1918–1919 (erroneously called the Spanish Flu), which is sometime referred to as the forgotten pandemic[3]. It earned this additional title, as there was a push to forget the catastrophic morbidity and loss, along with the lessons of what worked. In fact during the response to the COVID Pandemic (starting in 2020), some of the lessons from the SARS epidemic in 2003 (SARS-CoV, another coronavirus) were ignored early on.[4]

Emergency management agencies suffer from frequent turnover due to their smaller size and limited career prospects. Given this situation, it behooves these organizations to consolidate and sustain their institutional knowledge by integrating them through:

- Ensuring that systems and SOPs are developed and fully adopted into operating procedures, resisting the tendency to "reinvent the wheel".
- Integrating real-world experiences into training and exercises.
- Providing opportunities for staff at all levels to engage in hands-on emergency response through mutual aid deployments; this may also include other partners from the "Whole Community".
- Thorough briefings by experienced, departing staff members. Making sure that new staff know what plans, systems, and resources were used before for previous responses and their respective support functions, e.g. logistics and mass care, can help avoid the "reinventing the wheel" syndrome.

Creating and nurturing a learning culture that strives for continuous learning is the desired state in emergency management and public safety agencies, although it is difficult to maintain and sustain given the challenges and the great degree of public scrutiny they face.

NOTES

1 Deming, W. Edwards (2000). *The New Economics for Industry, Government, Education* (2nd ed.). MIT Press. ISBN 0-262-54116-5. W. E. Deming always referred to this as the Walter Shewhart Cycle for Continuous Learning and Improvement or the Plan-Do-Study-Act (PDSA). There are other works on which this based, such as the scientific method, design of experiments.
2 When collecting for a single data point and displaying that on a chart, we will often get a bell-shaped curve with an arithmetic mean (average). The deviation from the mean is called sigma. Three deviations on each side of the mean equal Six Sigma, which translates into 99.7% of the total data points that fall within those control limits (based on the deviations). In 10,000 data points that winds up being a total of 3.4 defects (fall outside the control limits).
3 Crosby, A. W. (2010). *America's for gotten Pandemic: The Influenza of 1918.* Cambridge University Press.
4 Davis IM. SARS-CoV: Lessons Learned; Opportunities Missed for SARS-CoV-2. Rev Med Virol. 2021 Jan;31(1):1–6. doi: 10.1002/rmv.2152. Epub 2020 Aug 18. PMID: 32808446; PMCID: PMC7460959; Fukami, Dr. Maki, Veasey, Frances (2022) *Did Lessons Learned from SARS Save Us from COVID-19? A Systems Thinking Perspective on International Case Studies,* International Institute of Global Resilience and ANSER. pp. 14–24.

Chapter 17

Project Leadership

Sometimes we frame this as if it's some sort of cliff that we go off at 3° Fahrenheit (1.5 Celsius increase). That's not what it is. It's a minefield, and we are walking farther and farther out onto that mine field. And the further we walk out in that mind field, the more danger we are going to encounter. So, at this point it's about limiting the danger; it's about limiting damage. We are not going to avoid dangerous climate change impacts because we're already seeing them... There is urgency, but there is agency. There is still time to take the actions necessary to avert truly catastrophic, global-scale climate consequences.

Michael Mann, distinguished Professor of atmospheric science at Penn State University, and the author of **The New Climate War: The Fight to Take Back Our Planet; People in the U.S. are suffering under intense heat warnings and heat advisories,** *July 25, 2022, Morning Edition, National Public Radio*

Leadership may be exhibited in many forms and at many levels in the fields of emergency and public safety management. When I introduce the topic of leadership in my university classes at John Jay and New York University (NYU), the first images that my students present are often public figures in politics, the military, or sports figures, mostly as these are obvious. However, increasingly I do see examples of their colleagues or immediate supervisors, often lower ranking public safety or first response managers, such as fire chiefs or a police lieutenant; these leaders see a difficult situation, where responsibility for tackling a critical problem does not clearly fall into an agency's or a role, and they take responsibility to work outside of official guidelines and rules to resolve the situation. This is defined as "leading without authority", which is similar to the authority assigned to project managers, especially at the junior level.

This chapter will discuss the importance of leadership in managing projects for emergency management and public safety; although some references and examples may be related, it is not intended to address leadership for emergency or crisis response. It will define leadership, discuss the different types of leadership, and identify the qualities of an effective leader and share some practical examples in emergency preparedness projects. The chapter will also discuss the challenges of leadership in emergency management and public safety, and how to overcome those challenges.

DOI: 10.1201/9781003201557-20

WHAT IS LEADERSHIP?

As discussed earlier, emergency and public safety agencies are increasingly facing an environment that is volatile, chaotic, and ambiguous. On top of this, they are also often faced with limited budgets and resources at their disposal, often working with a flat agency hierarchy and lean organizations, but are increasingly being asked to take on greater responsibilities outside of their regular mandates. With this often the case, they often have to rely on other agencies and partners to accomplish their mission.

This uncertain environment with limited resources and multi-agency coordination creates situations with high levels of complexity, where gaps may arise in meeting immediate critical needs, for example for mass care (emergency shelter, food, water) and security for special populations. At times, it can be unclear who has ownership of the problem. Leaders in emergency management, whether they are self-committed or voluntarily committed by their leaders, are typically the ones stepping into this breach. In the words of Warren

Table 17.1 Common Leadership Qualities

Leadership Quality	Definition and Traits/Behaviors Exhibited
Communication	Ability to communicate effectively with a wide range of groups, within the organization and external to the organization. While this includes foundational communication proficiency in written and oral delivery, to small or large groups, this also includes cultural competency.
Consensus and conflict management	Ability to create a consensus and negotiate through conflicts to achieve common goals.
Critical thinking	Ability to take in various sources of information and to critically evaluate it to produce a clear analysis. This includes breaking down an idea into constituent parts and understanding how they relate to one another.
Decision-making	Ability to analyze a situation, identify appropriate options, understand the consequences, and effect a decision.
Delegation	Transferring work, tasks, or projects to others in an effective manner that clearly communicates expectations for them to carry out the project.
Developing others	Working with others to define areas of improvement and helping them find ways to develop themselves, through coaching, training, mentoring, or simply providing feedback or demonstrating skills, behaviors, or other aspects of their work.
Financial management	Ability to financially manage an organization (or smaller unit) and balance budgets for fiscal solvency and peak operational performance.
Problem solving	Ability to identify and analyze problems and identify appropriate solutions. Oftentimes, this requires working with teams or other individuals in a cooperative manner.
Respect and concern	Genuine concern and respect for other people, their health and welfare, including their long-term goals and desires.
Self-management	Ability to regulate your efforts, time, and manage your emotions to convey a sense of control and equanimity. To effectively perform at your peak in order to get work done, without exhausting yourself or those around you.
Strategic thinking	Ability to define a vision, long-term goals, and understand the working environment to achieve that vision.
Team building	Ability to form and build a team and create a sense of joint ownership so that they can effectively accomplish the job, work, and/or mission.
Time management	Managing time prudently, scheduling your own work and work of others to maximize efficiency while balancing this with respect for others' schedules and capabilities.

Notes: The list of leadership qualities was researched and consolidated based on the following sources: Robert M. G. (former Secretary of Defense) (2016). *A Passion for Leadership: Lessons on Change and Reform from Fifty Years of Public Service*. Vintage Press; Willink, J., & Babin, L. (2015). *Extreme Ownership: How U.S. Navy SEALs Lead and Win*. St. Martin's Press; Covey, S. R. (2013). *The 7 Habits of Highly Effective People: Powerful Lessons in Personal Change*. Simon & Schuster; Hsieh, T. (2013). *Delivering Happiness: A Path to Profits, Passion, and Purpose*. Grand Central Publishing.

Bennis, an American author and leadership expert, "a leader is someone who does not abdicate responsibility, but seizes it".[1] In summary, leadership is the ability to influence others to achieve a common goal, and influence comprises the ability to motivate, inspire, and guide others to collaborate toward that goal. A common list of leadership qualities is given in Table 17.1.

We will explore these qualities and how they relate to managing projects in emergency and public safety preparedness and response further.

THE POWER DYNAMIC AND DIFFERENT TYPES OF LEADERSHIP

While leadership is the ability to influence others to act toward a common goal, power is the ability to compel or impel others to take action, whether it is toward a goal or not. Leadership includes the exercise of power, and different sources of power include[2] the following:

- **Legitimate power:** Power that is inherent from a person's position or title. For example, a director of a unit or department has legitimate power over their employees.
- **Expert power:** Power that comes from a person's knowledge or expertise. For example, a public health professional has expert power on public health decisions.
- **Referent power:** Power that comes from personal charisma, interpersonal skills, as well as the power of people with whom they are associated with. For example, a well-respected leader in a profession who is well connected has power and influence.
- **Reward/coercive power:** Power that comes from a person's ability to rewards or punish. Rewards include money, promotions, additional resources, or performance recognition. Punishments include taking away work privileges/benefits, bad performance reviews, demotion, reassignment, and suspension or firing someone from a job.

These sources of power are used with different leadership styles, and some of the most common types of leadership styles include the following:

- **Autocratic leadership:** Illustrated by a strong leader who tends to control the decision-making process and holds authority over the management of resources. Autocratic leaders tend to rely on legitimate and reward and coercive power.
- **Democratic leadership:** Demonstrated by a leader who consults with their immediate subordinates and partners before making decisions and setting objectives, collaborating to allocate and coordinate resources. Democratic leaders tend to rely on referent power, and a lesser degree on expert power.
- **Transformational leadership:** Characterized by a leader who inspires and motivates others to achieve objectives. Transformational leaders tend to rely on referent, and to a far lesser degree on expert power.
- **Servant-based leadership:** This leadership style is exhibited by a leader's focus on building a sense of teamwork and the development of the work community, serving the needs of the team and supporting their work by providing resources and problem-solving to help them meet customer/client needs.

CHALLENGES OF LEADERSHIP IN EMERGENCY MANAGEMENT AND PUBLIC SAFETY

Leadership in emergency management and public safety is challenging. Some of the major challenges include the following:

- **Elected leaders' expectations:** Elected political leaders may be less familiar with disaster situations and can have varying levels of expectations of their emergency and public safety agencies and understanding of their capabilities. While some elected leaders defer to experienced professionals and rely on their planning and recommendations, others may lack trust in moments of crisis and usurp the existing plans, going around the existing "Chain of Command", and the functional plans and channels to deliver critical goods and services, preferring to go through different agencies and using improvised plans, basically "reinventing the wheel". At the same time, emergency management and public safety agencies are being called on to take on responsibilities outside their normal management scope, e.g. immigration crisis, outreach and support for people experiencing homelessness, and the opioid crisis.
- **Public expectations:** With a wide diversity of public stakeholders that often have limited time and attention to major episodic incidents, there is frequently a limited understanding of the impacts they might face from disasters, what they might rely on from government agencies, and a lack of preparedness of the average resident. At times, there may be an unrealistic perception by members of the public that, even after being warned to prepare in advance or evacuate when called on to do so, the "cavalry will come over the hill" to their rescue when needs become critical. In addition, emergency management agencies are called on to meet the needs not only of the general public, but special needs populations: growing numbers of residents in assisted living facilities, non-English-speaking residents, people with mental and behavioral health challenges, including substance abuse.
- **Budget constraints:** It is a proverbial paradox; emergency management agencies ask the public taxpayers to spend money that taxpayers say they don't have, on something that may or may not occur, but, if it does occur, will be absolutely critical to meet their needs. Public budgets for emergency management and public safety are usually tight and certainly not growing. The irony is that we are witnessing even greater outlays on the costs of disaster recovery vs. mitigation and response to emergencies.
- **Small and lean organizations:** Except in a few jurisdictions, and to a lesser degree at the Federal and state levels, emergency management agencies and increasingly public safety agencies are lean organizations, working with limited resources, but when a major incident occurs, have to grow big and meet the capacity of handling support to community lifelines. In some counties emergency management agencies are led by a few professional staff, often with dual responsibilities, so time and attention is shared, spreading them thin. Marshaling and coordinating these resources escalates the leadership needs at all levels.
- **Staff turnover:** At the state and local levels, emergency management agencies experience a high level of turnover. In general, the smaller the agency, the greater the level of attrition. This is due to both limited career advancement, with young professionals moving into other areas of government, to state and Federal organizations, or to the

private sector, as well as burnout from frequent long hours and the stress of regular emergency response. This reduces the retention of institutional knowledge and experience with emergency and public safety agencies.

In addition to the challenges given above, in organizations where the emphasis is on command and control (vs. coordination and collaboration), with a more hierarchical approach, leadership at the top can be inelastic and not able to stretch to lower levels, at times limiting junior managers from taking proactive measures. Expanding leadership capabilities at all levels helps to increase overall organizational leadership capacity.

OPTIONS TO OVERCOME THE CHALLENGES OF LEADERSHIP

There are a number of ways to overcome the challenges of leadership in emergency management and public safety. Some of the most important ways include the following:

- **Assessment:** While it is often accepted that leadership trains are inherent in attitudes and values, leadership can be fostered in behaviors. Taking an inventory of leadership qualities for an individual using a leadership assessment can help identify areas to build skills and personal development.[3] Assessments can be used for long-term development and in team building with the aim of establishing leadership throughout the organization.[4]
- **Training:** This works in two ways for both leaders and emergency managers:
  - Leaders should receive training in emergency management and public safety, both in the public and private sector. Besides FEMA's Emergency Management Institute (EMI) and training at the state level, many emergency management organizations have begun establishing training programs.
  - For new emergency managers leadership should be included in the training curricula. This should include opportunities to practice this in the field and to bring real-life lessons learned into the classroom.
  - This type of training can help to develop the skills and knowledge needed to be effective leaders. Training and development is also offered through professional associations, such as International Association of Emergency Managers (IAEM) and the National Emergency Management Association (NEMA), as well as those in the private sector, such as Disaster Recovery Institute International (DRII) and Association of Continuity Professionals (ACP).
- **Mentoring and coaching:** While leadership lessons can often be learned as a natural consequence of managerial relationships, it is better to create an explicit and well-developed program to help develop future leaders in emergency and public safety organizations. Identify mentors and develop a formal leadership program, perhaps in concert with other related job competencies. Again, many professional organizations may offer mentoring opportunities as well.
- **Experience:** Leaders should gain experience in emergency management and public safety through their participation in after-action reviews, lessons learned following projects, and recording their successes and failures, as well as soliciting feedback from their peers and subordinates. This experience will help them to develop the skills and

judgment they need to make sound decisions under pressure, whether in blue, gray, or black sky conditions.

- **Community partnership and support:** Leaders need not only the support of their team, their organization, and key partners, but increasingly the private and non-profit sector, made up of CBOs/FBOs and K–12 and institutions of higher education. Fostering and cultivating a "whole of the community" alliance will help to share the challenges of leadership and achieve their goals.

Mission-Driven Leadership

Note – this is a first-person account from the author[5]

The weather was cold, but well above freezing, so I had hardly noticed the layer of black ice on the road. In a flash, I lost control of the Toyota Hilux I was driving, and it veered up onto the hillside suddenly, barrel rolling the vehicle onto its left side. The vehicle slid on its side for about 50 m and stayed on the road, luckily with no guard rail, not falling into the river on the left side embankment. I had my seatbelt on, but my two passengers, one to my right and one in back, did not and got knocked around. After checking to make sure that none of us were severely injured, I took the key out of the ignition, and we climbed out of the vehicle windows.

We were about an hour's drive from Tbilisi, the Capital of Georgia, with a damaged vehicle in an isolated part of the country. I had a meeting with the local Red Cross chapter, about another half-hour drive away. It was a miracle that none of us were injured, but that was not the only miracle that morning. About 5 min after we climbed out of the vehicle, a passenger car arrived. As we were blocking the road, and, by their nature, Georgians are a friendly people, with their help we managed to push the vehicle off its side and back onto its tires. The vehicle started up again when I turned the ignition key and was still drivable, miracle number two.

This was well before cell phone technology came into use, and each of our vehicles was equipped with a high frequency radio (Codan) and antenna attached to the front hood. The third miracle was that the radio was still working when I turned the engine. We called back to our headquarters in Tbilisi (Tango Base as we called it) to alert them about the accident. We also asked them to contact the Red Cross office and let them know we would be arriving late. Our headquarters office dispatched another vehicle to come to our aid and another driver, my driver.

You see, I was not supposed to be driving the vehicle. The driving conditions were not safe, and at the time, Georgia was still in the middle of a civil war. During the work week we were under strict orders to have our drivers at the wheel. That morning my driver was late, and I was impatient, so incautiously decided to depart for our destination without him, figuring I was used to driving in-country.

With this in mind, I was expecting to get "chewed out" by my supervisor, our Red Cross head of mission, Michael Stone. However, Michael did not admonish me. He was more concerned about making sure we were all safe and sound. The vehicle damage was secondary to him. Looking back at it, I believe he knew that I understood implicitly that the accident was punishment enough.

Michael was committed to the Red Cross' mission and led our sprawling operation of well over a couple of hundred staff with national operations reaching well over a million aid recipients. In our first meeting, over a cup of tea (ever the Brit) he told me about the principles of the Red Cross movement[6] and shared a book on the history of the Red Cross. During my two-year tenure with the International Federation of Red Cross and Red Crescent Societies (IFRC) in Georgia, I would become well acquainted with Michael's mission-driven leadership style, what might be called servant leadership in today's management parlance.

He was always even keeled and engaged his senior program administrators, myself included, in participatory management. He organized monthly meetings, alternating locations around the country where we had Red Cross offices, in which we would update the team on our progress, issues we were dealing with, and to propose new ideas and initiatives. These meetings involved all staff, expatriates and national staff and local Red Cross chapter leaders. They not only included the matters of business but also included sports activities (friendly, impromptu European football matches between offices) and dinners where we socialized. This instilled a strong sense of teamwork and esprit de corps.

While always the image of a British gentleman, Michael was never the effete aristocrat; he was always warm, good humored, and welcoming. Michael was always open to new initiatives and, if it aligned with our strategic objectives, he supported these projects. While we faced various challenges and might have our disagreements along the way, our team led by Michael understood that we had his unwavering support to meet our mission. His leadership lessons resonate with me to this day.

FINAL THOUGHTS

At all managerial levels, leadership is an essential element to build and maintain response capabilities in emergency management and public safety agencies. Of course, when responding to a significant incident, leaders in emergency and public safety management need to be competent to take into account the hazards they face and ensure the safety of their staff, partners, and the public. They also need to think quickly and make sound decisions under pressure, which often involves coordinating resources, collaborating and communicating effectively with others to build trust. In addition, leaders must be able to deal with the emotional and psychological toll that emergencies can take on their staff and the people served.

While this is true of response, it is equally important when it comes to establishing mitigation programs and physical measures to protect and reduce the impacts from hazards, such as flood, fire, wind, and seismic effects. Often the great work of project planners in prevention and mitigation requires adept leadership skills, by those who work in relative anonymity with persistence and tenacity, which often goes unrecognized as the negative impacts from the hazards and threats they address never wreak havoc.

NOTES

1 Bennis, W. G., & Nanus, B. (2007). *Leaders: Strategies for Taking Charge*. Harper Business.
2 Project Management Institute. (2017). *A Guide to the Project Management Body of Knowledge* (PMBOK guide). p. 63.

3 Canton, L. G. (2020). *Emergency Management: Concepts and Strategies for Effective Programs* (2nd ed.). John Wiley & Sons.

4 Canton, L. G. (2020). *Emergency Management: Concepts and Strategies for Effective Programs* (2nd ed.). John Wiley & Sons.

5 Mr. Michael Stone reviewed and agreed with the account and that his name can be used in this recollection.

6 This is Fundamental Principles of the International Red Cross and Red Crescent Movement, in place for over 50 years. See www.ifrc.org/who-we-are/international-red-cross-and-red-crescent-movement/fundamental-principles

Chapter 18

Next Steps

I have called this mental defect the Lucretius problem, after the Latin poetic philosopher who wrote that the fool believes that the tallest mountain in the world will be equal to the tallest one he has observed. We consider the biggest object of any kind that we have seen in our lives or heard about as the largest item that can possibly exist. And we have been doing this for millennia. In Pharaonic Egypt, which happens to be the first complete top-down nation-state managed by bureaucrats, scribes tracked the high-water mark of the Nile and used it as an estimate for a future worst-case scenario. The same can be seen in the Fukushima nuclear reactor, which experienced a catastrophic failure in 2011 when a tsunami struck. It had been built to withstand the worst past historical earthquake, with the builders not imagining much worse, and not thinking that the worst past event had to be a surprise, as it had no precedent.

Nassim Nicholas Taleb, **Antifragile: Things That Gain from Disorder**

I have been teaching and training in the areas of project and emergency management for 20 years, and when I finish with each lesson or class, the question that I always ask is: what did you learn today that you can apply tomorrow? and when it comes to the practical nature of emergency and public safety management, where contemplation may be a luxury I always ask: so what, now what? Now that we have covered how we might apply project management methodology in the areas of emergency management and public safety where do you go from here?

STARTING FROM WHERE YOU ARE

A journey of a thousand miles begins with a single step.

*Lao Tsu, the **Dao De Jing***

Much of how you might start implementing project management methods in your work largely depends on where you are starting from. If you are completely new to project management, then I recommend starting with the foundations: scope, schedule, and budget management; and then building out from there. If you are already using PM methods, then look at what method and techniques you might add or refine based on what was covered. Much like conducting any assessment, at an individual level you can do an inventory of project management competencies. Many of the project management competencies are also embodied in those for emergency management professionals, including the Next

DOI: 10.1201/9781003201557-21

Generation Core Competencies, such as critical thinking, continual learning, system literacy, and disaster risk management.[1]

GROWING A PROJECT CULTURE IN AN EMERGENCY MANAGEMENT ORGANIZATION

If you and your organization already manage projects, then it may benefit the practice to use the Project Management Maturity Model (PMMM),[2] described in Table 18.1. This is based on the Capability Maturity Model Integration (CMMI) which was developed by the Software Engineering Institute (SEI) at Carnegie Mellon University.[3]

The PMMM provides a structured approach for assessing an organization's project management capabilities and identifying areas where improvement may be needed. Each maturity level is a foundation for the next one, with the ultimate goal to reach Level 5, where ongoing process improvement has become part of the culture and continuously optimized. By following the model, organizations can help maintain and improve their project management practices and increase their capability to deliver successful projects.

As discussed earlier, the challenges that emergency management and public safety professionals face, often being pulled into dealing with the exigencies of the agency mission, make it a struggle to adopt and implement new management practices. I recommend implementing what you can within the resources and time you have. Even incremental improvements made at adopting basic practices, such as developing a scope of work with a standard scope document, can create greater clarity and organizational performance efficiencies.

PROJECT-BASED EXPERIENTIAL LEARNING

Some emergency management and public safety organizations have formed partnerships with academic programs in emergency management at universities and colleges to create

Table 18.1 Project Management Maturity Model

Level 1: Initial process	Project management processes are ad hoc and inconsistent. There is a lack of clear, consistent set of project management standards and procedures. Individual project managers largely achieve project success independently of the organization.
Level 2: Structured process and standards	Basic project management processes are established, and project successes can be repeated because the fundamental project management processes have been established, standardized, and documented.
Level 3: Organizational standards and institutionalized process	Organization-wide standards for project management are set. Processes are documented, standardized, and integrated into an organization-wide project management framework.
Level 4: Managed process	The organization monitors and controls its own project management processes through data collection and analysis. It has established project management metrics and uses them to monitor the quality of project management practices and predict project outcomes.
Level 5: Continuous improvement/optimizing	Project management best practices are shared across the organization, and the organization is committed to continuous improvement. The organization has processes in place to identify weaknesses and strengthen capabilities, often using statistical methods to predict performance and improve process effectiveness.

a pipeline of talent and gain access to the latest research and evidence-based practices for their organizations. These partnerships may include internships, service-learning projects, joint research projects, and opportunities for job-shadowing.

When it comes to establishing project-based learning initiatives, it is critical to ensure alignment between the strategic objectives of the client organization, academic program, and with the Next Generation of Core Competencies for Emergency Management Professionals.[4] Competency building for project management skills should focus not only on the core areas enumerated in this book, but, in particular, also on critical thinking, collaborative problem-solving, especially with multi-agencies and diverse partners, and in cultural competency.[5]

In order to be successful, project-based learning opportunities require a solid foundation with sustained engagement by program faculty and client sponsors, taking into account the potential challenges of legal, administrative, and scheduling issues.[6] A cornerstone of any project-based learning program would include a continuous learning loop, comprise intermittent debriefs, progress tracking, and reporting by the student, with direct feedback from the organization and academic advisor.

Emergency management, as an application-oriented field, benefits significantly from experiential learning, especially for students in undergraduate and graduate programs. This hands-on experience broadens their understanding, expands their professional network, and can demonstrate their career readiness. These opportunities hone critical thinking skills, aiding emergency managers in developing their competencies in handling complex situations and decisions with incomplete data. The learning opportunity is reciprocal; not only do students enhance their competencies by applying classroom-based knowledge in real-world scenarios, but this also expands the emergency management organizations perspective and expands their institutional knowledge-base.

PROFESSIONAL ORGANIZATIONS AND DEVELOPMENT

Another opportunity to pursue is to incorporate project management training and other learning opportunities at conferences, seminars, and symposia. I have been a member of the Project Management Institute (PMI) for over 20 years and a member of the International Association of Emergency Managers (IAEM) for 9 years, and other related professional associations, taking part in seminars, conferences, and workshops, at times delivering presentations and seminars. These are great opportunities to learn from others and reflect and enhance individual and organizational skills. When an organization sends one or more staff to a learning event, I encourage them to extend the learning by asking those staff members to share and present what they learned and how it might be used back in their own organization.

EVERY DAY IS GAME DAY

There is no shortage of disasters events, and at any given time in the world, we bear witness to natural hazards: a wildfire, major storm, flood, earthquake, or extreme heat and drought; or threats: cyberattacks, mass casualty shootings, and terrorist acts. Everyday

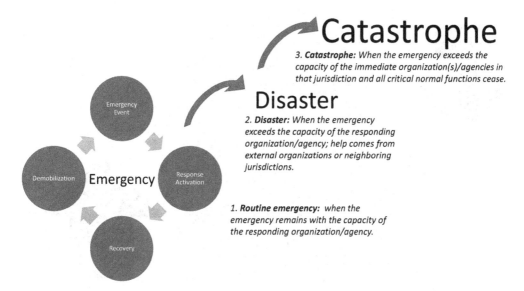

Figure 18.1 Scales of emergency response.

emergency and public safety professionals wake up and need to be prepared to respond to these types of events. As the saying goes, every day is game day.

As organizations and communities build and improve their capacities to respond, what once may have been a disaster or catastrophe can be handled in a more routine manner, as Figure 18.1 illustrates the different scales and levels of response. The increase in frequency and scale of impacts from a warming climate resulting in catastrophic events drives the need for these improved emergency response capacities at multiple levels.

Refer to Figure 18.2. Resilience through increased prevention, mitigation preparedness, as we see evident in the National Preparedness Goal and its related system and plans, emanates at all levels and helps address the ripple effects we see when hazards and threats impact our society, from the individual and household to the neighborhood and community, up through the cities and states and national levels. These are the building blocks that create resilience, as Figure 18.2 illustrates:

- A prepared household can help their neighbors;
- A prepared neighborhood can better help other neighborhoods and other communities;
- Prepared communities can aid other communities, cities, and towns;
- Through mutual aid (EMAC etc.) states can come to the aid of other states, territories, and tribal lands, and countries; the same is true for countries and NGOs.

We see frequent examples of this with Urban Search and Rescue (USAR) teams from throughout the country deploying to the scene of devastating earthquakes, Red Cross Disaster Assistance Teams standing up hot meals for disaster victims, and Incident Management Teams (IMTs) standing up to help relieve tired EOC staff to help coordinate emergency operations. I have seen this system at work up close. Does it work well? Yes, it

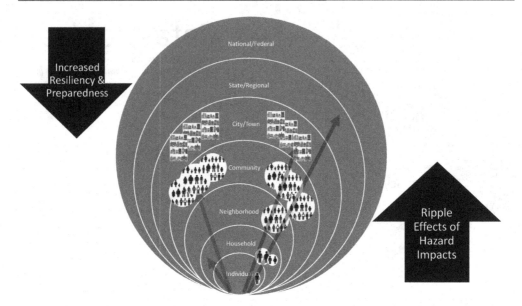

Figure 18.2 Resilience and preparedness levels vs. ripple effects of hazards.

does, most of the time. Can it use improvement? Of course, but as I have laid out in these final chapters, that is the mission of every agency involved in major emergency and public safety responses, building capacities and constant improvement.

RIDING THE WAVE

Among emergency managers and public safety professionals, there is a saying "never let a disaster go to waste", as the public and elected leaders are often open to investing to deal with the impacts of disaster situation once it has taken place and the immediate consequences are clear and costly. When funding is generally available and more easily accessible, we can look at this as "riding the wave" of funding, taking advantage of these opportunities: by putting in place prevention and mitigation measures; or to enhance the organization's capabilities, by planning, adding staffing, equipment, supplies, training, and exercises.

Refer to Figure 18.3. Investing in prevention and mitigation can reduce the need for preparedness and response, but these efforts will always be vital components of emergency management. While there will be disaster situations that overwhelm mitigation and preparedness levels, leaders in emergency management and public safety will continue to work tirelessly to ensure the safety and well-being of our communities. Whether forewarned or little to no notice is given, when major disaster events take place, leaders in emergency management and public safety, at all levels, navigate and ride through the waves of response, from the field to command posts to the EOCs, until the storm is over and calmer seas prevail.

The Rolling Wave Approach to Risk Management and Emergency Preparedness

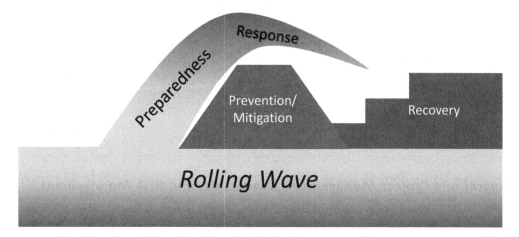

Figure 18.3 Rolling wave approach to emergency preparedness.

NOTES

1 Feldmann-Jensen, S., Jensen, S., & Maxwell Smith, S. (2019). The Next Generation Core Competencies for Emergency Management Professionals: Behavioral Anchors and Key Actions for Measurement.

2 Crawford, J. K. (2006) The Project Management Maturity Model Information Systems Management. 23(4): 50–58.
 DOI:10.1201/1078.10580530/46352.23.4.20060901/95113.7

3 Now managed by the Information Systems Audit and Control Association (ISACA) See: https://cmmiinstitute.com/

4 Feldmann-Jensen, S., Jensen, S., & Maxwell Smith, S. (2019). The Next Generation Core Competencies for Emergency Management Professionals: Behavioral Anchors and Key Actions for Measurement.

5 Carey, T. J. (2018). The Utilization of Client-Based Service-Learning in Emergency Management Graduate Curricula for the 21st century. Journal of Homeland Security Education, 7, 13–28.

6 Carey, T. J. (2018). The Utilization of Client-Based Service-Learning in Emergency Management Graduate Curricula for the 21st century. Journal of Homeland Security Education, 7, 13–28.

Bibliography

General and Project Management, Leadership, and Risk Management

Bennis, W. G., & Nanus, B. (2007). Leaders: Strategies for Taking Charge (Collins Business Essentials). Harper Business.

Clements, D. P. (2004). The 5% Rule: A Simple, Accurate Way to Estimate Project Completion Journal of Construction Engineering and Management, 130(2).

Committee of Sponsoring Organizations of the Treadway Commission. (2004). Appendix: COSO Enterprise Risk Management Framework. www.coso.org/guidance-erm

Covey, S. R. (2013). The 7 Habits of Highly Effective People: Powerful Lessons in Personal Change. Simon & Schuster.

Crawford, J. K. (2006). The Project Management Maturity Model. 23(4), 50–58, DOI:10.1201/ 1078.10580530/46352.23.4.20060901/95113.7

Deming, W. E. (2000). The New Economics for Industry, Government, Education. MIT Press.

Dinsmore, P. C., & Cabanis-Brewin, J. (2006). The AMA Handbook of Project Management. AMACOM.

Dumont, P. R., Gibson, G. E., & Fish, J. R. (1997). Scope Management Using Project Definition Rating Index. Journal of Management in Engineering, 13(5), 54–60.

Endsley, M. R. (1995). Toward a Theory of Situation Awareness in Dynamic Systems. Human Factors: The Journal of the Human Factors and Ergonomics Society, 37(1), 32–64. https://doi. org/10.1518/001872095779049543

Gates, R. M. (2016). A Passion for Leadership: Lessons on Change and Reform from Fifty Years of Public Service. Knopf.

Guzeviciute, G., & Varghese, D. (2023). Humanitarian Leadership for the Future: When 2023 Is the New 2030. International Federation of Red Cross and Red Crescent Societies (IFRC) Solferino Academy.

Heagney, J. (2022). Fundamentals of Project Management. HarperCollins Leadership.

Hillson, D. (September 2000). Project Risks: Identifying Causes, Risks, and Effects. PM Network, The Project Management Institute.

Hsieh, T. (2013). Delivering Happiness: A Path to Profits, Passion, and Purpose. Grand Central Publishing.

International Organization for Standardization. (2009). Risk Management-Principles and Guidelines. www.iso.org/standard/43170.html

Juran, J. M. (1974). Quality Control Handbook (3rd ed.). McGraw-Hill.

Kannan, R. (2014). Graphical Evaluation and Review Technique (GERT): The Panorama in the Computation and Visualization of Network-Based Project Management. Advances in Secure Computing, Internet Services, and Applications. IGI Global.

Katzenbach, J. R., & Smith, D. K. (2015). The Wisdom of Teams: Creating the High-Performance Organization. Harvard Business Review Press.

Kendrick, T. (2024). Identifying and Managing Project Risk: Essential Tools for Failure-Proofing Your Project (4th ed.). AMACOM.

Kerzner, H. (2022). Project Management: A Systems Approach to Planning, Scheduling, and Controlling (13th ed.). Wiley.

Larson, R. G., & Larson, E. N. (2009). Top Five Causes of Scope Creep … and What to Do about Them. Proceedings of the PMI Global Congress 2009—North America, Newtown Square, PA: Project Management Institute.

Lucas, H. C. (1981). Implementation: The Key to Successful Information Systems. Columbia University Press.

Malcolm, D. G., Roseboom, J. H., Clark, C. E., & Fazar, W. (1959). Application of a Technique for Research and Development Program Evaluation. Operations Research, 7(5), 646–669.

Maslow, A. H. (1943). A Theory of Human Motivation. Psychological Review, 50(4), 370–396. https://doi.org/10.1037/h0054346

Mehrabian, A. (1977). Nonverbal Communication. Aldine.

Moran, Dr. Alan (2014). Agile Risk Management and Scrum. Institute for Agile Risk Management.

Ng, M. H. Motivating Across Cultures: Encouraging Higher Performance 22 April 2021. Aperian Global.

Parkinson, C. N. (1955). Parkinson's Law. The Economist.

Project Management Institute. (2017a). A Guide to the Project Management Body of Knowledge (PMBOK Guide) (6th ed.). Project Management Institute.

Project Management Institute. (2017b). Agile Practice Guide (6th ed.). Project Management Institute.

Rose, K. (2014). Project Quality Management: Why, What and How. Cengage Learning.

Senor, D., & Singer, S. (2011). Start-Up Nation: The Story of Israel's Economic Miracle. Hachette Digital.

Stallings, R. A., & Quarantelli, E. L. (1985). Emergent Citizen Groups and Emergency Management. Public Administration Review, 45(2), 93–100.

Tuckman, B. W. (1965). Developmental Sequence in Small Groups. Psychological Bulletin, 63(6), 384–399.

Vogt, T. (1999). Likert Scale Response Options. Educational and Psychological Measurement, 59(3), 551–563.

Willink, J., & Babin, L. (2015). Extreme Ownership: How U.S. Navy SEALs Lead and Win. St. Martin's Press.

Witton F, Rasheed E. O., & Rotimi J. O. B. (2019). Does Leadership Style Differ between a Post-Disaster and Non-Disaster Response Project? A Study of Three Major Projects in New Zealand. Buildings, 9(9), 195. https://doi.org/10.3390/buildings9090195

Hazard Mitigation and Resilience

California Office of Emergency Services (CalOES). (2020, June). California Adaptation Planning Guide.

City of Los Angeles. (2018a). Resilient Los Angeles. Mayor's Office of Resilience, City of Los Angeles.

City of Los Angeles. (2018b). 2018 Local Hazard Mitigation Plan. Tetra Tech prepared for City of Los Angeles Emergency Management Department.

City of New York (2014). NYC's Risk Landscape: A Guide to Hazard Mitigation. NYC Emergency Management in Collaboration with the NYC Department of Planning and NYC Mayor's Office of Recovery and Resilience.

FEMA (2007). Multi-hazard Mitigation Planning Guidance under the Disaster Mitigation Act of 2000. DHS Federal Emergency Management Agency.

FEMA (2013). Local Mitigation Planning Handbook. DHS Federal Emergency Management Agency.

FEMA (2018). Threat and Hazard Identification and Risk Assessment (THIRA) and Stakeholder Preparedness Review Guide: Comprehensive Preparedness Guide (CPG) 201 (3rd ed.). DHS Federal Emergency Management Agency.

FEMA (2019). National Mitigation Investment Strategy. DHS Federal Emergency Management Agency.

Government of Ontario (2021). Incident Management System (IMS) Guidance Version 2.0. Office of the Chief, Emergency Management., Ministry of the Solicitor General, Government of Ontario Canada.

Multi-Hazard Mitigation Council. (2019). Natural Hazard Mitigation Saves: 2019 Report. Principal Investigator: Porter, K.; Co-Principal Investigators: Dash, N., Huyck, C., Santos, J., Scawthorn, C.; Investigators: Eguchi, M., Eguchi, R., Ghosh, S., Isteita, M., Mickey, K., Rashed, T., Reeder, A.; Schneider, P.; and Yuan, J., Directors, MMC. Investigator Intern: Cohen-Porter, A. National Institute of Building Sciences. Washington, DC. Retrieved from www.nibs.org

National Institute of Building Sciences. (2020). Mitigation Saves: Mitigation Saves Up to $13 per $1 Invested. Source: nibs.org/mitigationsaves.

Public Safety Canada (2011). National Emergency Response System. Operations Directorate, Public Safety Canada, Ottawa.

Texas General Land Office. Galveston's Bulwark Against the Sea: History of the Galveston Seawall. Save Texas History (Medium), Sept. 27, 2018. Source: https://savetexashistory.org/

United Nations (2015). Sendai Framework for Disaster Risk Reduction 2015–2030. Third UN World Conference on Disaster Risk Reduction in Sendai, Japan, in March 2015.

Emergency Management and Business Continuity

Aldrich, D. P. (2019). Black Wave: How Networks and Governance Shaped Japan's 3/11 Disasters. University of Chicago Press.

Bay Area Urban Area Security Initiative (UASI) FY 2020 – 2021 Annual Report. www.bayareauasi.org/sites/default/files/AR2020%20Final.pdf

Blanchard, B. W. (2005). Top Ten Competencies for Professional Emergency Management. Higher Education Program, Emergency Management Institute, Federal Emergency Management Agency. http://training.fema.gov/EMIWeb/edu/EMCompetencies.aspx

Boston Consulting Group (2015). UNICEF/WFP Return on Investment for Emergency Preparedness Study. UNICEF, World Food Program.

California Emergency Management Agency (2010). Foundation for the Standardized Emergency Management System (SEMS).

Canton, L. G. (2013). Disaster Planning and Management: Does One Leadership Style Work for Both?. Journal of Leadership Studies, 7(3), 47–50. https://doi.org/10.1002/jls.21297

Canton, L. G. (2020). Emergency Management: Concepts and Strategies for Effective Programs (2nd ed.). John Wiley & Sons.

Carey, T. J. (2018). The Utilization of Client-Based Service-Learning in Emergency Management Graduate Curricula for the 21st Century. Journal of Homeland Security Education, 7, 13–28.

Centers for Disease Control and Prevention (2018). Public Health Emergency Preparedness and Response Capabilities: National Standards for State, Local, Tribal, and Territorial Public Health. Centers for Disease Control and Prevention, US Department of Health and Human Services.

City of Los Angeles Emergency Management Department (2019). Training and Exercise Plan 2019–2022. Emergency Management Dept., City of Los Angeles.

Crosby, A. W. (2010). America's Forgotten Pandemic: The Influenza of 1918. Cambridge University Press.

Davis, I. M. (2020). SARS-CoV: Lessons Learned; Opportunities Missed for SARS-CoV-2. Rev Med Virol. 2021 Jan;31(1):1–6. doi: 10.1002/rmv.2152. Epub 2020 Aug. 18. PMID: 32808446; PMCID: PMC7460959

Davis, V. B. (2022). Lost and Turned Out: A Guide to Preparing Underserved Communities for Disasters. CreateSpace Independent Publishing Platform.

Emergency Management Accreditation Program (EMAP) (2019). Emergency Management Standard. EMAP.

Feldmann-Jensen, S., Jensen, S., & Maxwell Smith, S. (2019). The Next Generation Core Competencies for Emergency Management Professionals: Behavioral Anchors and Key Actions for Measurement.

FEMA (2010). Community Lifelines Implementation Toolkit: Comprehensive Information and Resources for Implementing Lifelines during Incident Response (v.2.0). DHS Federal Emergency Management Agency.

FEMA (2011). A Whole Community Approach to Emergency Management: Principles, Themes, and Pathways for Action. DHS Federal Emergency Management Agency.

FEMA (2014). FEMA Operational Planning Manual. DHS Federal Emergency Management Agency.

FEMA (2015). Incident Action Planning Guide (Rev.1). DHS Federal Emergency Management Agency

FEMA (2017). National Incident Management System. DHS Federal Emergency Management Agency.

FEMA (2020). Homeland Security Exercise and Evaluation Program (HSEEP). DHS Federal Emergency Management Agency.

FEMA (2021a). Developing and Maintaining Emergency Operations Plans: Comprehensive Preparedness Guide (CPG) 101 (V.3.0). DHS Federal Emergency Management Agency.

FEMA (2021b). Incident Complexity Guide: Planning, Preparedness, and Training. DHS Federal Emergency Management Agency.

Fretty, P. (2007). Disaster Strikes. PM Network, 21(2), 34–40.

Fukami, M., & Veasey, F. (2022). Did Lessons Learned from SARS Save Us from COVID-19? A Systems Thinking Perspective on International Case Studies. ANSER.

Haddow, G., Bullock, J., & Coppola, D. (2020). Emergency Management (7th ed.). Elsevier.

Hambridge, N. B., Howitt, A. M., & Giles, D. W. (2017, April). Coordination in Crises: Implementation of the National Incident Management System by Surface Transportation Agencies. Homeland Security Affairs, 13(2) (April 2017). www.hsaj.org/articles/13773

Inter-Agency Standing Committee (IASC) (2015a). Guidance: Cluster Coordination at Country Level. IASC Sub-Working Group on the Cluster Approach and the Global Cluster Coordinators' Group. UN OCHA.

Inter-Agency Standing Committee (IASC) (2015b). Reference Module: The Implementation of the Humanitarian Programme Cycle (v.2.0). Humanitarian Programme Cycle Steering Group. IASC, UN OCHA.

International Organization for Standardization (ISO) (2019). ISO 22301 Security and Resilience-Business Continuity Management Systems-Requirements (2nd ed.). ISO.

Jain, S. (2006). A Parametric Approach to Emergency Response Project Management. Paper Presented at PMI® Research Conference: New Directions in Project Management, Montréal, Québec, Canada. Newtown Square, PA: Project Management Institute.

Jamieson, G. (2005) NIMS and the Incident Command System. International Oil Spill Conference Proceedings. DOI:10.7901/2169-3358-2005-1-291

Knox, C.C., & Haupt, B. (2015). "Incorporating Cultural Competency Skills in Emergency Management Education", Disaster Prevention and Management, 24(5), pp. 619–634. https://doi.org/10.1108/DPM-04-2015-0089

Knox, C. C., & Haupt, B. (2020). Cultural Competency for Emergency and Crisis Management. Routledge.

Kyne, D., & Aldrich, D. P. (2019). Capturing Bonding, Bridging, and Linking Social Capital through Publicly Available Data. Risk, Hazards & Crisis in Public Policy, 10(4), 396–416.

LAFD Spokesperson: Nicholas Prange- Press Release. (January 13, 2020). LAFD Sees Record Low Structure Fire Fatality Deaths in 2019, Sets New Record for Number of Incident Response). Los Angeles Fire Department. https://www.lafd.org/news/lafd-sees-record-low-structure-fire-fatality-deaths-2019-sets-new-record-number-incident

Los Angeles Fire Department a Safer City 2.0. Strategic Plan 2018–2020. https://issuu.com/lafd/docs/strategic_plan_final_2018.02.09

McKinney, K. (2018). Moment of Truth: the Nature of Catastrophe and How to Prepare for Them. Savio Republic.

National Fire Protection Association. (2013). NFPA® 1600: Standard on Disaster/Emergency Management and Business Continuity Programs. NFPA.

National Weather Service. (2021, April 17). 10th Commemoration of Joplin, Missouri EF-5 Tornado. Retrieved from www.weather.gov/sgf/news_events_2011may22

Occupational Safety and Health Administration (OSHA). (2017, March). Personal Protective Equipment. OSHA 3151, March 2017 www.osha.gov/Publications/osha3151.pdf

Pierce, C. H. (1939). The Meteorological History of the New England Hurricane of Sept. 21, 1938.

Pine, J. C. (2017). Technology and Emergency Management (2nd ed.). Wiley.

Quarantelli, E. L. (1993). Technological and Natural Disasters and Ecological Problems: Similarities and Differences in Planning for and Managing Them. Disaster Research Center. University of Delaware.

Quarantelli, E. L. (1998). Major Criteria for Judging Disaster Planning and Managing their Applicability in Developing Countries. Preliminary Paper #268. Disaster Research Center. University of Delaware.

Red Alert: The Official Blog of the American Red Cross North Texas Region. American Red Cross North Texas Region website. https://redcrossntxblog.com/

Regional Catastrophic Planning Team of New York-New Jersey-Connecticut-Pennsylvania. (2015). Disabilities and Access and Functional Needs Emergency Operations Center Toolkit. RCPT NY-NJ-CT-PA.

Reynolds, B., & Seeger, M. W. (2014). Crisis and Emergency Risk Communication. US Department of Health and Human Services, Centers for Disease Control and Prevention.

Rubin, C. B. (2012). Emergency Management: The American Experience 1900–2010 (2nd ed.). CRC Press.

Schnurr, A. (2018). Black Sky Hazards: Systems Engineering as a Unique Tool to Prevent National Catastrophe. Journal of Systems Engineering, 31(2), 127–133.

Snedaker, S. (2007). Business Continuity and Disaster Recovery Planning for IT Professionals. Elsevier.

Tierney, K. J. (2019). Disasters: A Sociological Approach. Polity Press.

US Department of Homeland Security (2007). Target Capabilities List: A Companion to the National Preparedness Guidelines. US Department of Homeland Security.

Weisenfeld, P. E. (2011). Successes and Challenges of the Haiti Earthquake Response: The Experience of USAID. Emory International Law Review, 25(3), 1097–1120.

Wisconsin Council on Physical Disabilities (2019). Be Prepared, Have a Plan: Emergency Preparedness Toolkit for Persons with Disabilities. Source: cpd.wisconsin.gov/toolkit.htm.

Community and Disaster Recovery

Abramson, D., Culp, D., Johnson, L., & Bertman, L. (2013). Disaster Recovery: Guidance for Donors: Strategies for Helping Communities Heal. National Center for Disaster Preparedness, Columbia University Mailman School of Public Health.

American Red Cross. (2011). The Community Recovery Guide: A Practical Guide to Helping Communities Recover from Disaster.

Bau, M. S. (2006). Project Management for Post-disaster Events—Sri Lanka Project Galle 2005. Paper presented at PMI® Global Congress 2006—EMEA, Madrid, Spain. Newtown Square, PA: Project Management Institute.

Federal Emergency Management Agency (FEMA). (2011). The National Disaster Recovery Framework: Strengthening Disaster Recovery for the Nation. FEMA.

FEMA Federal Insurance and Mitigation Job Aid for Disaster Recovery Reform Act. (2019). www.fema.gov/sites/default/files/2020-07/fema_DRRA-1205-implementation-job-aid.pdf

Fister Gale, S. (2008). In for the Long Haul: Even after the Most Horrific Disasters, Attention Inevitably Fades; It's Up to Project Leaders to Keep Rebuilding Efforts Running. PM Network, 22(2), 32–39.

Gabriel, E. J. (2021). From Crisis to Recovery: Strategic Planning for Response, Resilience, and Recovery. John Wiley & Sons.

Hidayat, B., & Egbu, C. (2014). A Literature Review of the Role of Project Management in Post-Disaster Reconstruction. International Journal of Disaster Risk Reduction, 13, 130–141.

National Academies of Sciences, Engineering, and Medicine. (2015). Building Community Disaster Resilience through Private-Public Collaboration. The National Academies Press.

Phillips, B. D. (2022). Disaster Recovery: Principles and Practices. Business Expert Press.

Rodin, J. (2014). The Resilience Dividend: Being Strong in a World Where Things Go Wrong. Public Affairs.

Sterling, M. (2008). Project Management Methodology for Post Disaster Reconstruction. Paper presented at PMI® Global Congress 2008—Asia Pacific, Sydney, New South Wales, Australia. Newtown Square, PA: Project Management Institute.

Glossary

Note:
- Items with an asterisk (*) are from the *PMI Lexicon of Project Management Terms*. Source: www.pmi.org/pmbok-guide-standards/lexicon
- Items with a dagger (†) are from the FEMA Glossary or FEMA Course ICS 300 Intermediate Incident Command System for Expanding Incidents. Source: https://train ing.fema.gov/emiweb/is/icsresource/assets/glossary%20of%20related%20terms.pdf
- Items with a diamond (◊) are from the International Glossary of Resilience maintained by DRII. Source: https://drii.org/resources/viewglossary

Access and Functional Needs (also Disabilities, Access and Functional Needs; AFN or DAFN)†: Individual circumstances requiring assistance, accommodation, or modification for mobility, communication, transportation, safety, health maintenance, etc., due to any temporary or permanent situation that limits an individual's ability to take action in an emergency.

Activation◊: The implementation of emergency response and business continuity plans, procedures, and activities in response to a serious incident, emergency, event, or crisis.

Activity*: A distinct, scheduled portion of work performed during the course of a project.

ADA†: Americans with Disabilities Act, as amended 2008.

After-Action Review (AAR): A formal feedback process on project performance from all key stakeholders to gain from lessons learned, and includes a facilitation method, either through a written survey, in-person debrief, or preferably both.

Agency†: A government element with a specific function offering a particular kind of assistance.

Agile Project Management: An iterative and incremental approach to developing projects in shorter cycles that focuses on finished units of work that can be used to benefit the end-user. The process starts with a "burndown" list of user requirements and focused on a rigorous, dynamic development process vs. a deterministic planning process. Agile methods are designed to adapt to end-user requirements, with user-testing and feedback at the end of each development cycle, with the capability to prioritize meeting end-users' expectations as opposed to requirements fully planned out in advance.

Alert◊: Notification that a potential disaster situation is imminent, exists, or has occurred; usually includes a directive for personnel to stand by for possible activation.

All-Hazards†: A classification encompassing all conditions, environmental or human-caused, that have the potential to cause injury, illness, or death; damage to or loss of equipment, infrastructure services, or property; or alternatively causing functional degradation to social, economic, or environmental aspects. These include accidents, technological events, natural disasters, space weather, domestic- and foreign-sponsored terrorist attacks, acts of war, weapons of mass destruction, and chemical, biological (including pandemic), radiological, nuclear, or explosive events.

Analogous Estimating*: A technique for estimating the duration or cost of an activity or a project using historical data from a similar activity or project. See also *bottom-up estimating, parametric estimating, program evaluation and review technique (PERT),* and *three-point estimating.*

Assumption*: A factor in the planning process considered to be true, real, or certain, without proof or demonstration.

Audit†: A financial review undertaken internally by a separate unit solely responsible for audits within the organization by an independent public accounting firm.

Authority Having Jurisdiction (AHJ)†: An entity that has the authority and responsibility for developing, implementing, maintaining, and overseeing the qualification process within its organization or jurisdiction. This may be a state or Federal agency, training commission, NGO, private sector company, or a tribal or local agency such as a police, fire, or public works department. In some cases, the AHJ may provide support to multiple disciplines that collaborate as a part of a team (e.g. an IMT).

Backup◊: A process by which data, electronic or paper based, is copied in some form so as to be available and used if the original data from which it originated is lost, destroyed, or corrupted.

Benchmarking◊: Comparing a benchmark with a baseline or with best practice. The term benchmarking is also used to mean creating a series of benchmarks over time and comparing the results to measure progress or improvement.

Business Continuity◊: An ongoing process to ensure that the necessary steps are taken to identify the impact of potential losses and maintain viable recovery strategies, recovery plans, and continuity of services.

Business Continuity Management◊: Holistic management process that identifies potential threats to an organization and the impacts to business operations those threats, if realized, might cause, and which provides a framework for building organizational resilience with the capability of an effective response that safeguards the interests of its key stakeholders, reputation, brand, and value-creating activities.

Business Continuity Plan (BCP)◊: A documented collection of procedures and information that is developed, compiled, and maintained in readiness for use in an incident to enable an organization to continue to deliver its critical products and services at an acceptable predefined level.

Business Continuity Program◊: Ongoing management and governance process supported by top management and appropriately resourced to implement and maintain business continuity management.

Business Impact Analysis (BIA)◊: A method of identifying the effects of failing to perform a function or requirement.

Capacity Building: Efforts aimed to develop human skills or societal infrastructure within a community or organization needed to respond, reduce the level of risk or the effects of a disaster. (From the U.N. International Strategy for Disaster Reduction.)

Catastrophic Incident†: Any natural or manmade incident, including terrorism that results in extraordinary levels of mass casualties, damage, or disruption severely affecting the population, infrastructure, environment, economy, national morale, and/or government functions. A catastrophic event could result in sustained national impacts over a prolonged period of time; almost immediately exceeds resources normally available to local, state, tribal, and private sector authorities in the impacted area; and significantly interrupts governmental operations and emergency services to such an extent that national security could be threatened.

CBRNE†: Chemical, biological, radiological, neurological, and explosive weapons. An example of CBRNE-specific equipment is a monitor. A CBRNE-specific pharmaceutical is an item such as an autoinjector.

Chain of Command†: The orderly line of authority within the ranks of incident management organizations.

Change Control*: A process whereby modifications to documents, deliverables, or baselines associated with the project are identified, documented, approved, or rejected.

Change Management◊: Change management refers to the broad processes for managing organizational change. Change management encompasses planning, oversight or governance, project management, testing, and implementation.

Change Request*: A formal proposal to modify a document, deliverable, or baseline.

Chief†: The ICS title for individuals responsible for the management of functional sections: Operations, Planning, Logistics, and Finance/Administration.

Command†: The act of directing, ordering, or controlling by virtue of explicit statutory, regulatory, or delegated authority.

Command Center◊: The location, local to the event but outside the immediate affected area, where tactical response, recovery, and restoration activities are managed. There could be more than one command center for each event reporting to a single emergency operations center.

Communications Management Plan*: A component of the project, program, or portfolio management plan that describes how, when, and by whom information will be administered and disseminated. See also *project management plan*.

Community†: A network of individuals and families, businesses, governmental and non-governmental organizations, and other civic organizations that reside or operate within a shared geographical boundary and may be represented by a common political leadership at a regional, county, municipal, or neighborhood level.

Command Center◊: The location, local to the event but outside the immediate affected area, where tactical response, recovery, and restoration activities are managed. There could be more than one command center for each event reporting to a single emergency operations center.

Community Resilience†: The ability to prepare for anticipated hazards, adapt to changing conditions, and withstand and recover rapidly from disruptions. Activities such as disaster preparedness (prevention, protection, mitigation, response, and recovery) are key steps to resilience.

Component (portfolio component): A discrete element of a portfolio that is a program project or other work element related to the achievement of the portfolio strategic objectives.[1]

Constraint*: A factor that limits the options for managing a project, program, portfolio, or process.

Contingency Plan*: A document describing actions that the project team can take if predetermined trigger conditions occur.

Contingency Reserve*: Time or money allocated in the schedule or cost baseline for known risks with active response strategies.

Continuity◊: Strategic and tactical capability, pre-approved by management, of an organization to plan for and respond to conditions, situations, and events in order to continue operations at an acceptable predefined level.

Continuity of Government (COG)†: A coordinated effort within the executive, legislative, or judicial branches to ensure that essential functions continue to be performed before, during, and after an emergency or threat. Continuity of government is intended to preserve the statutory and constitutional authority of elected officials at all levels of government across the United States.

Continuity of Operations (COOP) Plan (or Continuity Plan; NIST SP 800-34)◊: A predetermined set of instructions or procedures that describe how an organization's mission-essential functions will be sustained within 12 h and for up to 30 days as a result of a disaster event before returning to normal operations.

Cooperating Agency†: An agency supplying assistance other than direct operational or support functions or resources to the incident management effort.

Core Capability†: An element defined in the National Preparedness Goal as necessary to prevent, protect against, mitigate, respond to, and recover from the threats and hazards that pose the greatest risk.

Corrective Action*: An intentional activity that realigns the performance of the project work with the project management plan.

Cost-Benefit Analysis†: A quantitative procedure that assesses the desirability of a project by taking a view of avoided future damages compared to the cost of a project. The outcome of the analysis is a cost-benefit ratio, which demonstrates whether the net present value of benefits exceeds the net present value of costs.

Cost Management Plan*: A component of a project or program management plan that describes how costs will be planned, structured, and controlled.

Crisis◊: A critical event, which, if not handled in an appropriate manner, may dramatically impact an organization's profitability, reputation, or ability to operate. Or, an occurrence and/or perception that threatens the operations, staff, shareholder value, stakeholders, brand, reputation, trust, and/or strategic/business goals of an organization.

Crisis Management◊: The overall coordination of an organization's response to a crisis, in an effective, timely manner, with the goal of avoiding or minimizing damage to the organization's profitability, reputation, and ability to operate. Crisis management is the term most commonly used in for-profit and non-profit sphere vs. emergency management in the governmental sphere.

Critical Infrastructure†: Assets, systems, and networks, whether physical or virtual, so vital to the United States that the incapacitation or destruction of such assets, systems, or networks would have a debilitating impact on security, national economic security, national public health or safety, or any combination of those matters. Critical infrastructure includes any system or asset that, if disabled or disrupted in any significant way, would result in catastrophic loss of life or catastrophic economic loss. Some examples of critical infrastructure include: public water systems serving large population centers; primary data storage and processing facilities, stock exchanges or major banking centers; chemical facilities located in close proximity to large population centers; major power generation facilities exceeding 2,000 MW and supporting the regional electric grid; hydroelectric facilities and dams producing power in excess of 2,000 MW that could cause catastrophic loss of life if breached; nuclear power plants; major underground gas, water, phone, and electrical supplies affecting a large population.

Critical Path*: The sequence of activities that represents the longest path through a project, which determines the shortest possible duration. Critical Path Method is used to estimate the minimum project duration and determine the amount of scheduling flexibility on the logical network paths within the schedule model.

Cyber Attack (or Cyberattack or Cyber-Attack)◊: An attempt to damage, disrupt, or gain unauthorized access to a computer, computer system, or electronic communications network. An attack, via cyberspace, targeting an enterprise's use of cyberspace for the purpose of disrupting, disabling, destroying, or maliciously controlling a computing environment/infrastructure; or destroying the integrity of the data or stealing controlled information.

Cyber Resilience◊: An entity's ability to continuously deliver their products and services despite any adverse cyber events by actively protecting against known or potential threats; planning for the recoverability of applications and data; adapting to changing threat landscapes; effectively training personnel about the existing threats; and ensuring that response plans are maintained and exercised.

Cybersecurity◊: The prevention of damage to, unauthorized use of, exploitation of, and – if needed – the restoration of electronic information and communications systems, and the information they contain, in order to strengthen the confidentiality, integrity, and availability of these systems.

Decomposition*: A technique used for dividing and subdividing the project scope and project deliverables into smaller, more manageable parts.

Delegation of Authority†: A statement that the agency executive delegating authority and assigning responsibility provides to the Incident Commander. The delegation of authority can include priorities, expectations, constraints, and other considerations or guidelines, as needed. Identification, by position, of the authorities for making policy determinations and decisions at senior levels, field levels, and all other organizational locations. Generally, predetermined delegations of authority will take effect when normal channels of direction have been disrupted and will lapse when these channels have been reestablished. (Federal Continuity Directive-1.)

Deliverable*: Any unique and verifiable product, result, or capability to perform a service that is produced to complete a process, phase, or project.

Demobilization†: The orderly, safe, and efficient return of an incident resource to its original location and status.

Department Operations Center (DOC or DEOC) †: An operations or coordination center dedicated to a single, specific department or agency. The focus of a DOC is on internal agency incident management and response. DOCs are often linked to and/or physically represented in a combined agency EOC by an authorized agent(s) for the department or agency.

Disaster◊: A sudden, unplanned catastrophic event causing unacceptable damage or loss. An event that compromises an organization's ability to provide critical functions, processes, or services for some unacceptable period of time and where an organization's management invokes their recovery plans.

Disaster Recovery (DR)◊: The technical aspect of business continuity. The collection of resources and activities to reestablish information technology services (including components such as infrastructure, telecommunications, systems, applications, and data) at an alternate site following a disruption of IT services. Disaster recovery includes subsequent resumption and restoration of those operations at a more permanent site.

Disaster Recovery Plan◊: A written plan for recovering one or more information systems at an alternate facility in response to a major hardware or software failure or destruction of facilities.

Disaster Risk Reduction (UN International Strategy for Disaster Reduction): The concept and practice of reducing disaster risks through systematic efforts to analyze and manage the causal factors of disasters, including through reduced exposure to hazards, lessened vulnerability of people and property, wise management of land and the environment, and improved preparedness for adverse events. A comprehensive approach to reduce disaster risks is set out in the United Nations-endorsed Hyogo Framework for Action, adopted in 2005, whose expected outcome is "The substantial reduction of disaster losses, in lives and the social, economic and environmental assets of communities and countries".

Disruption◊: An event that interrupts normal business, functions, operations, or processes, whether anticipated (e.g. hurricane, political unrest) or unanticipated (e.g. a blackout, terror attack, technology failure, or earthquake).

DMAIC: A six sigma approach to managing quality that includes the following steps: define the problem, measure the problem, analyze the problem and develop the solution, implement the solution, and control for the results.

Downtime◊: A period in time when something is not in operation.

Duration*: The total number of work periods required to complete an activity (also called a work package) or total project duration. Periods may be expressed in different increments of time: hours, days, weeks, months, etc. Activity duration may include both non-working and working time, as opposed to activity effort.

Economies of Scale: The cost advantages experienced when the level of production is increased. This occurs due to the spread fixed costs over a larger number of units produced, leveraging bulk buying discounts, and more efficient use of resources at larger scales of production.

Economies of Scope: The cost advantages achieved when multiple-related products or services are produced together, rather than individually. The average total cost of production decreases as a result of increasing the number of different goods or services produced due to shared resources such as financial control, marketing and sales networks, complementary goods and services, and diversification of risk.

Effort*: The number of labor units required to complete a schedule activity or work breakdown structure component, often expressed in hours, days, or weeks.

Emergency†: Any incident, whether natural, technological, or human-caused, that necessitates responsive action to protect life or property.

Emergency Action Plan: An Emergency Action Plan describes the organizational emergency response that is incident or site specific. It consists of security and safety guidance and procedures, such as evacuations and medical first aid in response to hazard-specific impacts such as fire, flood, high-wind, earthquake, and threats. In contrast, the Emergency Operations Plan describes how an organization will execute its emergency response to a variety of emergency incidents and scales.

Emergency Management†: Emergency management is the managerial function charged with creating the framework within which communities reduce vulnerability to hazards and cope with disasters. Emergency management seeks to promote safer, less vulnerable communities with the capacity to cope with hazards and disasters. Emergency Management protects communities by coordinating and integrating all activities necessary to build, sustain, and improve the capability to mitigate against, prepare for, respond to, and recover from threatened or actual natural disasters, acts of terrorism, or other manmade disasters.

Emergency Management Assistance Compact (EMAC)†: A congressionally ratified agreement that provides form and structure to interstate mutual aid. Through EMAC, a disaster-affected state can request and receive assistance from other member states quickly and efficiently, resolving two key issues up front: liability and reimbursement.

Emergency Medical Services (EMS)†: The Emergency Medical Service (EMS) system is responsible for providing pre-hospital (or out-of-hospital) care by paramedics, emergency medical technicians (EMTs), and medical first responders. The goal of EMS is to provide early treatment to those in need of urgent medical care and ultimately rapid transportation to a facility providing more definitive treatment.

Emergency Operations Center (EOC)†: The physical location where the coordination of information and resources to support incident management (on-scene operations) activities normally takes place. An EOC may be a temporary facility or located in a more central or permanently established facility, perhaps at a higher level of organization within a jurisdiction.

Emergency Operations Plan (EOP)†: A plan for responding to a variety of potential hazards.

Emergency Preparedness◊: The capability that enables an organization or community to respond to an emergency in a coordinated, timely, and effective manner to prevent the loss of life and minimize injury and property damage.

Emergency Response◊: The immediate reaction and response to an emergency situation commonly focusing on ensuring life-safety and reducing the severity of the incident.

Emergency Response Team (ERT)◊: Qualified and authorized personnel who have been trained to provide immediate assistance and coordinate resources to prevent the loss of life and minimize injury and property damage.

Emergency Response Plan◊: A documented plan usually addressing the immediate reaction and response to an emergency situation. This term is commonly used in for-profit and non-profit organizations in place of an Emergency Operations Plan.

Emergency Support Function (ESF)†: The grouping of governmental and certain private sector capabilities into an organizational structure to provide capabilities and services most likely needed to manage domestic incidents. The grouping of governmental and certain private sector capabilities into an organizational structure to provide capabilities and services most likely needed to manage domestic incidents.

Enterprise Risk Management (ERM)◊: ERM includes the methods and processes used by organizations to manage risks and seize opportunities related to the achievement of their objectives. ERM provides a framework for risk management, which typically involves identifying particular events or circumstances relevant to the organization's objectives (risks and opportunities), assessing them in terms of likelihood and magnitude of impact, determining a response strategy, and monitoring progress. By identifying and proactively addressing risks and opportunities, business enterprises protect and create value for their stakeholders, including owners, employees, customers, regulators, and society overall.

Essential Functions◊: The critical activities performed by organizations, especially after a disruption of normal activities.

Essential Services◊: Infrastructure services without which a building or area would be considered disabled and unable to provide normal operating services; typically includes utilities (water, gas, electricity, telecommunications); and may also include standby power systems or environmental control systems.

Exercise◊: An activity in which an organization's plan(s) is practiced in part or in whole to ensure that the plan(s) contains the appropriate guidance, procedures, actions, resources, and trained staff and produces the desired outcome when put into effect. Exercises can be Table-Top Exercises (TTX), Drills, Functional, and Full-Scale Exercises. See *HSEEP*.

Federal Emergency Management Agency (FEMA)†: An agency within the U.S. Department of Homeland Security charged with preparing the nation (USA) to responding to disasters. FEMA administers the National flood Insurance program (NFIP) and supports local efforts on the state, tribal, territorial, and local level in a presidentially declared disaster.

Finance/Administration Section†: The ICS Section responsible for an incident's administrative and financial considerations.

Flood Map†: A Flood Insurance Rate Map (FIRM), Flood Boundary and Floodway Map (FBFM), and Flood Hazard Boundary Map (FHBM) are all flood maps produced by FEMA. The FIRM is the most common type of map and most communities have this type of map. Recent flood map products include digital FIRMs, which are created using digital methods and can be incorporated into a community's Geographic Information System (GIS).

Flood Zones†: Flood hazard areas identified on the Flood Insurance Rate Map are identified as a Special Flood Hazard Area (SFHA). SFHA is defined as the area that will be inundated by the flood event having a 1% chance of being equaled or exceeded in any given year.

Gantt Chart*: A bar chart of schedule information where activities are listed on the vertical axis, dates are shown on the horizontal axis, and activity durations are shown as horizontal bars placed according to start and finish.

Hazard: Something that is potentially dangerous or harmful, often the root cause of an unwanted outcome†. A dangerous phenomenon, substance, human activity, or condition that may cause loss of life, injury, or other health impacts, property damage, loss of livelihoods and services, social and economic disruption, or environmental damage◊.

Hazard Mitigation†: Any action taken to reduce or eliminate the long-term risk to human life and property from hazards.

HIPAA†: Health Insurance Portability and Accountability Act regulates how medically related personally identifiable information (data) is managed and secured by organizations.

Homeland Security Exercise and Evaluation Program (HSEEP) †: A US Department of Homeland Security (DHS) program that provides a set of guiding principles for exercise programs, as well as a common approach to exercise program management, design, development, conduct, evaluation, and improvement planning.

Impact◊: The adverse effect of an event on an organization on a community. Impacts are usually described as financial and non-financial and are further divided into specific types of impact.

Impact Analysis◊: Process of analyzing all operational functions and the effect that an operational interruption might have upon them. Impact analysis includes business impact analysis, the identification of critical business assets, functions, processes, and resources as well as an evaluation of the potential damage or loss that may be caused to the organization resulting from a disruption (or a change in the business or operating environment).

Incident†◊: An occurrence, natural or manmade, that has the potential to cause interruption, disruption, loss, emergency, crisis, disaster, or catastrophe and necessitates a response to protect life or property. In NIMS, the word "incident" includes planned events as well as emergencies and/or disasters of all kinds and sizes.

Incident Action Plan (IAP; sometimes referred to as an Incident Management Plan)†: An oral or written plan containing the objectives established by the Incident Commander or Unified Command and addressing tactics and support activities for the planned operational period, generally 12–24 h.

Incident Command (IC)†: The ICS organizational element responsible for overall management of the incident and consisting of the Incident Commander or Unified Command and any additional Command Staff activated.

Incident Command Post (ICP)†: The field location where the primary functions of incident command are performed. The ICP may be co-located with the Incident Base or other incident facilities.

Incident Command System (ICS)†: A standardized approach to the command, control, and coordination of on-scene incident management, providing a common hierarchy within which personnel from multiple organizations can be effective. ICS is the combination of procedures, personnel, facilities, equipment, and communications operating within a common organizational structure, designed to aid in the management of on-scene resources during incidents. It is used for all kinds of incidents and is applicable to small, as well as large and complex, incidents, including planned events.

Incident Commander†: The individual responsible for on-scene incident activities, including developing incident objectives and ordering and releasing resources. The Incident Commander has overall authority and responsibility for conducting incident operations.

Incident Management†: The broad spectrum of activities and organizations providing operations, coordination, and support applied at all levels of government, using both governmental and non-governmental resources to plan for, respond to, and recover from an incident, regardless of cause, size, or complexity.

Incident Management Team (IMT)†: A rostered group of ICS-qualified personnel consisting of an Incident Commander, Command and General Staff, and personnel assigned to other key ICS positions.

Incident Response◊: The response of an organization to a disaster or other significant event that may significantly impact the organization, its people, or its ability to function productively. An incident response may include evacuation of a facility, initiating a disaster recovery plan, performing a damage assessment, and any other measures necessary to bring an organization to a more stable status.

Incident Objective†: A statement of an outcome to be accomplished or achieved. Incident objectives are used to select strategies and tactics. Incident objectives should be realistic, achievable, and measurable, yet flexible enough to allow strategic and tactical alternatives.

Indirect Costs: Costs that have been incurred for managing and supporting the activities of a grant-funded project that are not directly incurred by the performance of the grant activities and include such cost areas as depreciation or use allowances on buildings and equipment, costs of operating and maintaining facilities, general administration and other general expenses such as the salaries and expenses of executive officers, personnel administration, and accounting. Indirect costs are typically expressed as an Indirect Cost Rate.

Interoperability†: The ability of systems, personnel, and equipment to provide and receive functionality, data, information, and/or services to and from other systems, personnel, and equipment, between both public and private agencies, departments, and other organizations, in a manner enabling them to operate effectively together.

Joint Field Office (JFO)†: A temporary Federal multi-agency coordination center established locally to facilitate field-level domestic incident management activities and provides a central location for coordination of Federal, state, local, tribal, non-governmental, and private-sector organizations with primary responsibility for activities associated with threat response and incident support.

Jurisdiction†: Jurisdiction has two definitions depending on the context: A range or sphere of authority. Public agencies have jurisdiction at an incident related to their legal responsibilities and authority. Jurisdictional authority at an incident can be political or geographical (e.g. local, state, tribal, territorial, and Federal boundary lines) and/or functional (e.g. law enforcement, public health); or a political subdivision (e.g. municipality, county, parish, state, Federal) with the responsibility for ensuring public safety, health, and welfare within its legal authorities and geographic boundaries.

Just-in-Time (JIT)◊: System whereby dependencies for critical business processes are provided exactly when required, without requiring intermediate inventory. JIT Training is training that is provided to emergency management staff (EOC, Command Center, or Field Staff) at the time of deployment.

Lessons Learned*: The knowledge gained during a project which shows how project events were addressed or should be addressed in the future for the purpose of improving future performance.

Level of Effort*: An activity that does not produce definitive end-products and is measured by the passage of time. [Note: Level of effort is one of three earned value management (EVM) types of activities used to measure work performance.]

Likelihood◊: Chance of something happening, whether defined, measured, or estimated objectively or subjectively. It can use general descriptors (such as rare, unlikely, likely, almost certain), frequencies, or mathematical probabilities. It can be expressed qualitatively or quantitatively. See *Probability*.

Local Government†: Public entities responsible for the security and welfare of a designated area as established by law. A county, municipality, city, town, township, local public authority, school district, special district, intrastate district, council of governments (regardless of whether the council of governments is incorporated as a non-profit corporation under state law), regional or interstate government entity, or agency or instrumentality of a local government; a tribe or authorized tribal entity, or in Alaska, a Native Village or Alaska Regional Native Corporation; a rural community, unincorporated town or village, or other public entity.

Logical Relationship*: A dependency between two activities or between an activity and a milestone. Dependency relationships between tasks may be finish-to-finish, finish-to-start, start-to-finish, and start-to-start.

Logistics†: The process and procedure for providing resources and other services to support incident management.

Long-Term Recovery†: Phase of recovery that may continue for months or years and addresses complete redevelopment and revitalization of the impacted area, rebuilding or relocating damaged or destroyed social, economic, natural, and built environments and a move to self-sufficiency, sustainability, and resilience.

Major Disaster†: Any natural catastrophe (including any hurricane, tornado, storm, high water, wind-driven water, tidal wave, tsunami, earthquake, volcanic eruption, landslide, mudslide, snowstorm, or drought), or, regardless of cause, any fire, flood, or explosion, in any part of the United States, which in the determination of the President causes damage of sufficient severity and magnitude to warrant major disaster assistance to supplement the efforts and available resources of states, local governments, and disaster relief organizations in alleviating the damage, loss, hardship, or suffering caused thereby.

Matrix Organization*: An organizational structure in which the project manager shares authority with the functional manager temporarily to assign work and apply resources.

Maximum Tolerable Downtime (MTD)◊: The amount of time mission/business process can be disrupted without causing significant harm to the organization's mission.

Milestone*: A significant point or event in a project, program, or portfolio. Milestones on a project schedule usually have zero duration.

Milestone Schedule*: A type of schedule that presents milestones with planned dates.

Mission Area†: One of five areas (prevention, protection, mitigation, response, and recovery) designated in the National Preparedness Goal to group core capabilities.

Mitigation†: The capabilities necessary to reduce the loss of life and property from natural and/or manmade disasters by lessening the impacts of disasters. Mitigation capabilities include, but are not limited to, community-wide risk reduction projects; efforts to improve the resilience of critical infrastructure and key resource lifelines; risk reduction for specific vulnerabilities from natural hazards or acts of terrorism; and initiatives to reduce future risks after a disaster has occurred.

Mitigation Activity†: A mitigation measure, project, plan, or action proposed to reduce risk of future damage, hardship, loss, or suffering from disasters.

Mobilization†: The processes and procedures for activating, assembling, and transporting resources that have been requested to respond to or support an incident.

Mutual Aid or Mutual-Aid/Assistance Agreement†: A written or oral agreement between and among agencies/organizations and/or jurisdictions that provides a mechanism to quickly obtain assistance in the form of personnel, equipment, materials, and other associated services. The primary objective is to facilitate the rapid, short-term deployment of support prior to, during, and/or after an incident.

National Flood Insurance Program (NFIP)†: A program that makes federally backed flood insurance available in those states and communities that agree to adopt and enforce flood-plain management ordinances to reduce future flood damage.

National Incident Management System (NIMS)†: A systematic, proactive approach to guide all levels of government, NGOs, and the private sector to work together to prevent, protect against, mitigate, respond to, and recover from the effects of incidents. NIMS provides stakeholders across the whole community with the shared vocabulary, systems, and processes to successfully deliver the capabilities described in the National Preparedness System. NIMS provides a consistent foundation for dealing with all incidents, ranging from daily occurrences to incidents requiring a coordinated Federal response.

National Response Framework (NRF)†: NRF is part of the National Strategy for Homeland Security that presents the guiding principles enabling all levels of domestic response partners to prepare for and provide a unified national response to disasters and emergencies. Building on the existing National Incident Management System (NIMS) as well as Incident Command System (ICS) standardization, the NRF's coordinating structures are always in effect for implementation at any level and at any time for local, state, and national emergency or disaster response.

Natural Hazard (UN International Strategy for Disaster Reduction): Natural process or phenomenon that may cause loss of life, injury, or other health impacts, property damage, loss of livelihoods and services, social and economic disruption, or environmental damage.

Network Logic*: All activity dependencies in a project schedule network diagram. See also *Logical Relationship*.

Network Path*: A sequence of activities connected by logical relationships in a project schedule network diagram. See also *Network Logic* and *Logical Relationship*.

Non-governmental Organization (NGO)†: A group that is based on the interests of its members, individuals, or institutions. An NGO is not created by a government, but it may work cooperatively with government. Examples of NGOs include faith-based groups, relief agencies, organizations that support people with access and functional needs, and animal welfare organizations. This term is generally used for international humanitarian and social service organizations.

Non-profit Organization†: A tax-exempt organization that serves the public interest. In general, the purpose of this type of organization must be charitable, educational, scientific, religious, or literary. It does not declare a profit and utilizes all revenue available after normal operating expenses in service to the public interest. This organization is a 501(c)(3) or a 501(c)(4) designate.

Operational Period†: The time scheduled for executing a given set of operation actions, as specified in the IAP. Operational periods can be of various lengths but are typically 12–24 h.

Operational Work (Operations): Operational work is the routine, ongoing work of the organization such as production and/or service delivery. It is often defined by a set of standard operating procedures or set processes.

Operations (Emergency)†: Operations ensure the efficient and effective delivery of immediate emergency assistance to individuals and communities impacted by major disasters, emergencies, or acts of terrorism.

Operations Section†: The ICS Section responsible for implementing tactical incident operations described in the IAP. In ICS, the Operations Section may include subordinate branches, divisions, and/or groups.

Opportunity*: A risk that would have a positive effect on one or more project objectives.

Pandemic: An epidemic or infectious disease that has spread across a large region, for instance multiple continents or worldwide, affecting a substantial number of individuals.

Parametric Estimating*: An estimating technique in which an algorithm is used to calculate cost or duration based on historical data and project parameters. See also *analogous estimating, bottom-up estimating, program evaluation and review technique (PERT),* and *three-point estimating.*

Personal Protective Equipment (PPE)[2]: PPE is equipment worn to minimize exposure to hazards that cause serious workplace injuries and illnesses. These injuries and illnesses may result from contact with chemical, radiological, physical, electrical, mechanical, or other workplace hazards. Personal protective equipment may include items such as gloves, safety glasses and shoes, earplugs or muffs, hard hats, respirators, or coveralls, vests, and full body suits.

Phase Gate*: A review at the end of a phase in which a decision is made to continue to the next phase, to continue with modification, or to end a project or program.

Portfolio: Projects, programs, subsidiary portfolios, and operations managed as a group to achieve strategic objectives. See also *program* and *project.*

Planning Section†: The ICS Section that collects, evaluates, and disseminates operational information related to the incident and for the preparation and documentation of the IAP. This section also maintains information on the current and forecasted situation and on the status of resources assigned to the incident.

Points of Distribution (POD)†: A POD is a centralized locations in an impacted area where survivors pick up life-sustaining relief supplies following a disaster or emergency. PODs most often distribute food commodities to disaster survivors. POD is also used to refer to Points of Dispensing when they distribute vaccines or antibiotics to infectious disease outbreaks or for a biohazard or bioweapon attack.

Preparedness†: Actions taken to plan, organize, equip, train, and exercise to build and sustain the capabilities necessary to prevent, protect against, mitigate the effects of, respond to and recover from threats and hazards. Sometimes called readiness. See *Readiness.*

Prevention†: The capabilities necessary to avoid, prevent, or stop a threatened or actual act of terrorism. In national preparedness guidance, the term "prevention" refers to preventing imminent threats.

Private Sector†: Organizations and individuals that are not part of any governmental structure. The private sector includes for-profit and not-for-profit organizations, formal and informal structures, commerce, and industry.

Probability (ISO): Expression of the chance (likelihood) that a considered event will take place. See *Likelihood.*

Procurement Management Plan*: A component of the project or program management plan that describes how a team will acquire goods and services from outside of the performing organization.

Product Life Cycle*: The series of phases that represent the evolution of a product, from concept through delivery, growth, maturity, and to retirement. See also *project life cycle.*

Program*: Related projects, subsidiary programs, and program activities managed in a coordinated manner to obtain benefits not available from managing them individually.

Program Management*: The application of knowledge, skills, and principles to a program to achieve the program objectives and to obtain benefits and control not available by managing program components individually. See also *portfolio management* and *project management.*

Program Manager*: The person authorized by the performing organization to lead the team or teams responsible for achieving program objectives. See also *portfolio manager* and *project manager.*

Progressive Elaboration*: The iterative process of increasing the level of detail in a project management plan as greater amounts of information and more accurate estimates become available.

Project*: An undertaking involving the commitment of a significant investment of resources (people, equipment, material/supplies, services, systems, etc.) to produce a unique outcome (product, service, or result) in a specified amount of time.

Project Management Body of Knowledge (PMBOK©): A standard framework with a set of guidelines, better practices, and terminology for project management, developed by the Project Management Institute (PMI).

Project Scope*: The work performed to deliver a product, service, or result with the specified features and functions.

Project Scope Statement*: The description of the project scope, major deliverables, assumptions, and constraints.

Protection†: The capabilities necessary to secure the homeland against acts of terrorism and manmade or natural disasters.

Public Assistance†: The Public Assistance (PA) Grant Program under FEMA provides assistance to state, local, tribal, and territorial governments, and certain types of private non-profit (PNP) organizations so that communities can quickly respond to and recover from major disasters or emergencies declared by the President.

Public Information†: Processes, procedures, and systems for communicating timely, accurate, and accessible information on an incident's cause, size, and current situation; resources committed; and other matters of general interest to the public, responders, and additional stakeholders (both directly affected and indirectly affected).

Quality Management Plan*: A component of the project or program management plan that describes how an organization's policies, procedures, and guidelines will be implemented to achieve the quality objectives.

Readiness◊: Activities implemented prior to an incident that may be used to support and enhance mitigation of, response to, and recovery from disruptions. Sometimes called preparedness. See *Preparedness*.

Recovery†: The capabilities necessary to assist communities affected by an incident to recover effectively, including, but not limited to, rebuilding infrastructure systems; providing adequate interim and long-term housing for survivors; restoring health, social, and community services; promoting economic development; and restoring natural and cultural resources.

Recovery Point Objective (RPO; ISO 22301): Point to which information used by an activity must be restored to enable the activity to operate on resumption. This also referred to as "maximum data loss".

Recovery Time Objective (RTO)◊: Time goal for the restoration and recovery of functions or resources based on the acceptable down time and acceptable level of performance in case of a disruption of operations.

Residual Risk◊: The level of risk remaining after all cost-effective actions have been taken to lessen the impact, probability, and consequences of a specific risk or group of risks, subject to an organization's risk appetite.

Response†: The capabilities necessary to save lives, protect property and the environment, and meet basic human needs after an incident has occurred.

Requirements Management Plan*: A component of the project or program management plan that describes how requirements will be analyzed, documented, and managed.

Resilience: The ability for a community or organization to return to a reasonable level of its previous normal level and type of activities (e.g. back to business, daily routines) within an acceptable period of time. The ability to prepare for and adapt to changing conditions and recover rapidly from operational disruptions; this includes the ability to withstand and recover from deliberate attacks, accidents, or naturally occurring threats or incidents◊.

Resource Management◊: A system for identifying available resources to enable timely access needed to prevent, mitigate, prepare for, respond to, maintain continuity during, or recover from an incident.

Restoration◊: Process of planning for and/or implementing procedures for the repair of hardware and systems, recuperation of staff, relocation of the primary site and its contents, and returning to normal operations at the permanent operational location.

Return on Investment (ROI)◊: A measurement of the expected benefit returned on an investment. In the simplest financial sense, it is the net profit of an investment divided by the net worth of the assets invested.

Risk: The potential for an unwanted outcome resulting from an incident, event, or occurrence, as determined by its likelihood and the associated consequences†. An uncertain event or condition that, if it occurs, has a positive or negative effect on one or more project objectives*. See also *Opportunity* and *Threat*.

Risk Acceptance◊: A management decision to take no action to mitigate the impact of a particular risk.

Risk Analysis: A methodical evaluation of the components and characteristics of risk event, usually involving a quantitative look, reviewing causal factors and immediate impacts and consequences, for the purpose of risk treatment (for emergencies this may be referred to as a "course of action") to reduce the probability and impact of a risk event.

Risk Appetite◊: Total amount of risk that an organization is prepared to accept, tolerate, or be exposed to at any point in time.

Risk Assessment: A formal risk assessment consists of evaluating the probability and impact of a risk event for the purpose of prioritizing and deciding on whether it warrants further analysis and treatment. It is typically an informal assessment, such as basic study of flooding frequency and potential losses.

Risk Avoidance◊: An informed decision to not become involved in or to withdraw from a risk situation.

Risk Management: Coordinated activities to direct and control an organization with regard to risk[3] (ISO 31000:2009). The identification, assessment, and response to risk to a specific objective[4] (COSO ERM:2004). The identification, assessment, and response to risk to a specific objective. Strategic risk management is a business discipline that drives deliberation and action regarding uncertainties and untapped opportunities that affect an organization's strategy and strategy execution[5] (The Risk Management Society).

Risk Management Plan*: A component of the project, program, or portfolio management plan that describes how risk management activities will be structured and performed.

Risk Reduction◊: A selective application of appropriate techniques and management principles to reduce either probability of an occurrence or its impact, or both.

Risk Tolerance◊: Organization's readiness to bear the risk after risk treatments in order to achieve its objectives.

Risk Transfer◊: A common technique used by risk managers to address or mitigate potential exposures of the organization. A series of techniques describing the various means of addressing risk through insurance and similar products.

Root Cause Analysis (RCA)◊: An activity that identifies the root cause of an incident or problem.

Scenario◊: A predefined set of conditions that describe an emergency event for planning purposes or to help support the exercises or training, with the ultimate intention to develop and prepare resources for an incident response strategy.

Schedule Management Plan*: A component of the project or program management plan that establishes the criteria and the activities for developing, monitoring, and controlling the schedule.

Scope Creep*: The uncontrolled expansion to product or project scope without adjustments to time, cost, and resources.

Scope of Work: A document that describes all of the work to be performed on a project.

Scope Management Plan*: A component of the project or program management plan that describes how the scope will be defined, developed, monitored, controlled, and validated. See also *project management plan*.

Service Animals†: Any guide dog, signal dog, assistive dog, seizure dog, or other animal individually trained to do work or perform tasks for the benefit of an individual with a disability, including but not limited to guiding individuals with impaired vision, alerting individuals with impaired hearing to intruders or sounds, providing minimal protection or rescue work, pulling a wheelchair, or fetching dropped items.

Service Level Agreement (SLA)◊: A formal agreement between a service provider (whether internal or external) and their client (whether internal or external), which covers the nature, quality, availability, scope, and response of the service provider. The SLA should cover day-to-day situations and disaster situations, as the need for the service may vary in a disaster.

Shelter†: A place of refuge that provides life-sustaining services in a congregate facility for individuals who have been displaced by an emergency or a disaster.

Sheltering†: Housing that provides short-term refuge and life-sustaining services for disaster survivors who have been displaced from their homes and are unable to meet their own immediate post-disaster housing needs.

Situation Report (SitRep)†: Confirmed or verified information regarding the specific details relating to an incident.

Situational Awareness[6]: The conscious knowledge of the immediate environment and the events that are occurring in it. It involves perception, comprehension, and projection of the elements in the environment, and how they relate to one another. Situational awareness is a planning concept used for data collection and information management for effective decision-making in emergency incidents.

SLTT†: State, local territorial, and tribal. A term used by FEMA in the context of emergency management to define different levels of government and their roles and responsibilities for disaster response, recovery, preparedness, and mitigation, often in coordination with federal agencies like FEMA, and how Federal funding for any prevention, mitigation, and preparedness grant programs/projects, response or disaster assistance is coordinated.

Social Services: Social services are a range of public services provided by governmental and/or non-profit organizations that are designed to support the well-being of individuals, by providing protection and assistance to help improve the social and psychological functioning and families in a society. This is targeted to people who are most vulnerable due to factors such as poverty, unemployment, illness, disability, discrimination, or old age. Social services are intended to improve and maintain their quality of life and ability to participate fully in society.

Special Medical Needs: Requirements of individuals who have a specific health condition or disability that necessitates additional assistance, services, or accommodations. These needs could be for medical care, long-term health maintenance, or accommodations to perform activities of daily living. In a disaster or emergency situation, this can also refer to individuals who may need additional assistance because they rely on medical treatments or durable medical equipment, like dialysis machine or oxygen concentrator, which can be disrupted in these scenarios.

Special Needs Populations†: Members of a society who may have additional needs before, during, and after an incident in functional areas, including but not limited to: maintaining independence, communication, transportation, supervision, and medical care. Individuals in need of additional response assistance may include those who have disabilities, live in institutionalized settings, are elderly, are children, are from diverse cultures, have limited English proficiency or are non-English speaking, or are transportation disadvantaged.

Sponsor*: An individual or a group that provides resources and support for the project, program, or portfolio and is accountable for enabling success. See also *stakeholder*.

Staffing Management Plan*: A component of the resource management plan that describes when and how team members will be acquired and how long they will be needed.

Stakeholder*: An individual, group, or organization that may affect, be affected by, or perceive itself to be affected by a decision, activity, or outcome of a project, program, or portfolio. See also *sponsor*.

Stakeholder Engagement Plan*: A component of the project or program management plan that identifies the strategies and actions required to promote productive involvement of stakeholders in project or program decision-making and execution. See also *project management plan*.

Standard Operating Procedure (SOP)†: A reference document or an operations manual that provides the purpose, authorities, duration, and details for the preferred method of performing a single function or several interrelated functions in a uniform manner.

Subject Matter Expert (SME): Someone with expertise in a given subject matter area that provides technical guidance for decision-making on approaches, designs, and/or project plans.

Succession Plan◊: A predetermined plan for ensuring the continuity of authority, decision-making, and communication in the event that key members of executive management unexpectedly become incapacitated.

Supply Chain◊: The linked processes that begins with the acquisition of raw material and extends through the delivery of products or services to the end-user across the modes of transport. The supply chain may include suppliers, vendors, manufacturing facilities, logistics providers, internal distribution centers, distributors, wholesalers, and other entities that lead to the end-user.

Sustainability†: Meeting the needs of the present without compromising the ability of future generations to meet their own needs.

Task (also called a Work Package)*: The work defined at the lowest level of the work breakdown structure for which cost and duration are estimated and managed.

Task Force (TF)†: Any combination of resources of different kinds and/or types assembled to support a specific mission or operational need.

Terrorism†: Any activity that involves an act that is dangerous to human life or potentially destructive of critical infrastructure and is a violation of the criminal laws of the United States or of any state or other subdivision of the United States; and appears to be intended to intimidate or coerce a civilian population, or to influence the policy of a government by intimidation or coercion, or to affect the conduct of a government by mass destruction, assassination, or kidnapping.

THIRA†: Threat and Hazard Identification and Risk Assessment. THIRA is a formalized methodology for community risk assessment as described in the FEMA Comprehensive Preparedness Guide (CPG) 201.

Threat: A risk that would have a negative effect on one or more project objectives. See also *opportunity* and *risk**. Natural or manmade occurrence, individual, entity, or action that has or indicates the potential to harm life, information, operations, the environment, and/or property†.

Trigger◊: An event that causes risk to occur or a system to initiate an emergency response or deployment of a continuity and recovery plan.

Triple Constraint*: Primary constraints of time, cost (budget), and scope (work performed on the project; this may sometimes include quality).

Underserved Populations/Communities†: Groups that have limited or no access to resources or that are otherwise disenfranchised. These groups may include people who are socioeconomically disadvantaged; people with limited English proficiency; geographically isolated or educationally disenfranchised people; people of color as well as those of ethnic and national origin minorities; women and children; individuals with disabilities and others with access and functional needs; and seniors.

Urban Area Security Initiative (UASI)†: A federally funded program designed to assist jurisdictions considered at high risk for incidents involving weapons of mass destruction.

USAR†: Federally sponsored and sanctioned teams designated for urban search and rescue. USAR teams are specially trained for search and rescue following heavily damaged structures, in the aftermath of earthquakes, tornados, hurricanes, or other disaster events which cause severe structural damage.

Vulnerability◊: The degree to which a person, asset, process, information, infrastructure, or other resources are exposed to the actions or effects of a risk, event or other negative occurrence.

Vulnerability Assessment◊: Systematic examination of an information system, supply chain, process, assets, and resources to determine the adequacy of security and safety measures, identify deficiencies, provide data from which to predict the effectiveness of proposed protection and mitigation measures, and confirm the adequacy of such measures after implementation.

Walk-Through◊: A thorough demonstration or explanation that details each step of a process.

Whole Community†: A focus on enabling the participation in incident management activities of a wide range of players from the private and non-profit sectors, including NGOs and the general public, in conjunction with the participation of all levels of government, to foster better coordination and working relationships. The whole community is an inclusive approach to preparedness and calls for the involvement of everyone – not just the government – in preparedness efforts and presents it as a shared responsibility. By working together, everyone can help keep the nation safe from harm and help keep it resilient when struck by hazards, such as natural disasters, acts of terrorism, and pandemics.

Wildland Urban Interface†: The zone of transition between unoccupied land and human development. It is the line, area, or zone where structures and other human development meet or intermingle with undeveloped wildland or vegetative fuels.

Work Breakdown Structure (WBS)*: A hierarchical decomposition of the total scope of work to be carried out by the project team to accomplish the project objectives and create the required deliverables.

Workaround*: An immediate and temporary response to an issue, for which a prior response had not been planned or was not effective.

NOTES

1 Project Management Institute. (2017). The Standard for Portfolio Management, Fourth Edition, pp. 115

2 Occupational Safety and Health Administration (OSHA), Personal Protective Equipment. See: www.osha.gov/personal-protective-equipment

3 International Organization for Standardization, "Risk Management-Principles and Guidelines," ISO 2009 (Geneva, Switzerland: International Organization for Standardization, 2009), p. 2.

4 Committee of Sponsoring Organizations of the Treadway Commission, "Appendix: COSO Enterprise Risk Management Framework," 2004, www.erm.coso.org, p. 4 (accessed February 22, 2012)

5 Risk and Insurance Management Society (2019), "RIMS Defines an Emerging Discipline

6 Endsley, M.R. (1995b). "Toward a theory of situation awareness in dynamic systems". *Human Factors*. 37 (1): 32–64. doi:10.1518/001872095779049543. S2CID 8347993.

Index

Printed in the United States
by Baker & Taylor Publisher Services